Reviewers praise *The*

"The world respects Francis as a bri breakthrough discoveries benefiting mankind. For a decade i have been privileged to admire him as a devoted family man and talented musician with a charmingly sharp wit. This intellectually honest, spiritually grounded reconciliation of God and science helps answer your greatest questions. I was profoundly enlightened and believe this important book should be required reading."

—Naomi Judd

"Francis Collins, one of the world's most distinguished scientists, treats the relationship of science and religion with reason and reverence. Collins's mix of clear technical exposition and personal reflection is infused with an intellectual and spiritual honesty. Everyone who questions how religious faith could be reconciled with scientific knowledge, everyone who fears that modern science attacks the heart of religious faith, everyone interested in an enlightened discussion of a crucial issue of our time should read this book."

—William D. Phillips, 1997 Nobel Laureate in Physics

"*The Language of God* is a powerful confession of belief from one of the world's leading scientists. Refuting the tired stereotypes of hostility between science and religion, Francis Collins challenges his readers to find a unity of knowledge that encompasses both faith and reason. Faith, as he demonstrates, is not the enemy of scientific rationality, but its perfect complement. This powerful and personal testament from the Director of the Human Genome Project will surprise some, delight others, and will make a lasting contribution to the great culture of human understanding."

—Kenneth Miller, Brown University, author of *Finding Darwin's God*

"Francis S. Collins proves that there is a place for apologetics. He presents, in a surprisingly easy-to-read manner, scientific validation for a worldview in which God is not only present, but actively at work."

—Tony Campolo, Eastern University, author of
*Speaking My Mind: The Radical Evangelical Prophet
Tackles Tough Issues Christians Are Afraid to Face*

"Collins has a wonderful way of explaining complex genetic topics so that they seem almost self-evident. His love of science and his love of God come through very clearly. . . . The author is a brilliant scientist who didn't park his intelligence at the curb in order to believe."

—*JAMA, the Journal of the American Medical Association*

"Timely and incisive. Collins shows how our understanding of evolution, far from standing in the way of faith, reveals a universe of ever greater ingenuity and subtlety."

—Paul Davies, author of *The Fifth Miracle: The Search for the Origin and Meaning of Life*

"Dr. Collins, the leader of one of history's greatest scientific achievements, is also a man of profound faith. In this superb book, he . . . brings reason and reconciliation to several issues that currently divide our culture. I could not put the book down."

—Dr. Armand Nicholi, author of *The Question of God*

"A remarkable book, in which one of the world's leading geneticists shares his passionate love of science and his story of personal faith. Compelling reading for anyone reflecting on the relation of science and faith."

—Alister E. McGrath, author of *Dawkins' God: Genes, Memes and the Meaning of Life*

"Dr. Francis S. Collins is making an enormous contribution toward helping people resolve their confusion over conflicts between science and faith. As a seeker after truth, Dr. Collins has discovered that faith and science are not only compatible but complementary. Dr. Collins is another 'pencil in the hand of God' to bring understanding and reconciliation in a field of conflict."

—Douglas E. Coe, Washington, D.C.

"Most believers don't care to listen to an atheistic scientist calling the idea of God a mythology created to explain what humans don't understand, and academic atheists are just as uninterested in scientific lectures from Bible literalists. Collins, however, has both the standing and the desire to promote a third way. . . . Collins insists on overlaying and intertwining [science and faith]."

—*Time*

"*The Language of God* is a book of enormous value. At a time when so many people on both sides are trying to foment a conflict between science and religion, Collins is a sorely needed voice of reason. His book may do more to promote better understanding between the worlds of faith and science than any other so far written."

—*First Things*

"Collins wants to convince the religious that science leaves plenty of leeway for belief in God. And he wants scientists to know that religious faith can actually be based on reason. Religion and science can not only coexist, but they can reinforce one another, each contributing to a sense of awe and wonder at God. . . . He is an intelligent, honest guide to the problems posed by science and religion, and watching him wrestle with his material is instructive."

—*The Globe and Mail* (Toronto)

"Collins challenges some religious fundamentalists and atheists who insist that the world of instruments and measurements cannot be reconciled with faith. [*The Language of God*] is part tutorial, conceived mostly for young people of faith struggling to reconcile the religious with the scientific. . . . But it is also a guide to how, from [Collins's] viewpoint, science allows plenty of room for faith."

—*The News & Observer* (Raleigh, N.C.)

"This book does more than just review the voluminous evidence for evolution. . . . Collins's aims are broader, more ambassadorial. To Collins, evolution and faith are altogether compatible."

—*The Wilson Quarterly*

"No one knows better than Francis Collins how easy it might be for scientists to play God. He is perhaps the world's foremost geneticist, a man who has led efforts to decode human DNA, along the way developing a revolutionary method of screening genes for disease. . . . In his provocative new book, *The Language of God,* he melds his Christian faith and rational empiricism in a way few world renowned scientists would dare."

—*The Express* (London)

"In these days when people separate themselves into camps of 'us and them' with dismaying ease, it is remarkable for someone with a foot in each camp to tell both sides they should get along. Remarkable. Necessary. And extremely welcome."

—*The Arizona Republic*

"Besides offering a lovely, impassioned, and transparently sincere defense of his own Christian faith, Collins argues that one need not choose between Darwin and God . . . a strong and moving case for religious belief."

—*The Weekly Standard*

"[Collins's] argument that science and faith are compatible deserves a wide hearing. It lets non-churchgoers, too, consider spiritual questions without feeling awkward. The French scientist Pierre Simon Laplace, when asked about God, told Napoleon, 'I have no need of that hypothesis.' It's hard to share his view after reading this book."

—*The New York Times Book Review*

"[Collins] has written well for a general audience. The facts of nature are laid out clearly. His religious life is as well, and that makes the book rare if not unique."

—*Science*

"Compelling. Dr. Collins has superb credentials for tackling this sticky subject. . . . Collins adeptly translates arcane science into easily understood terms. Wise and timely."

.—*National Catholic Reporter*

THE LANGUAGE OF GOD

A Scientist Presents Evidence for Belief

FRANCIS S. COLLINS

Free Press
New York London Toronto Sydney

FREE PRESS
A Division of Simon & Schuster, Inc.
1230 Avenue of the Americas
New York, NY 10020

First Free Press trade paperback edition July 2007

FREE PRESS and colophon are trademarks of Simon & Schuster, Inc.

For information about special discounts for bulk purchases,
please contact Simon & Schuster Special Sales at 1-800-456-6798
or business@simonandschuster.com

Designed by Davina Mock

Figure 5.1 (right side) taken from *Darwin* by Niles Eldredge (W. W.
Norton, New York, 2005). All other line drawings: Michael Hagelberg.
Excerpt on page 31 from "The Gap" from *A Severe Mercy* by Sheldon
Vanauken. Copyright © 1977, 1980 by Sheldon Vanauken. Reprinted
by permission of HarperCollins Publishers.

Manufactured in the United States of America

20 19 18 17 16 15 14 13 12

The Library of Congress has catalogued the hardcover edition as follows:
Collins, Francis S.
The language of God: a scientist presents evidence for belief/
Francis S. Collins.
p. cm.
Includes bibliographical references and index.
1. Religion and science. 2. Apologetics. I. Title.
BL 240.3 .C66   2006
215—dc22   2006045316
ISBN-13: 978-0-7432-8639-8
ISBN-10:   0-7432-8639-1
ISBN-13: 978-1-4165-4274-2 (pbk)
ISBN-10:   1-4165-4274-4 (pbk)

To my parents, who taught me to love learning.

CONTENTS

ix

Contents

INTRODUCTION

O N A WARM SUMMER DAY just six months into the new millennium, humankind crossed a bridge into a momentous new era. An announcement beamed around the world, highlighted in virtually all major newspapers, trumpeted that the first draft of the human genome, our own instruction book, had been assembled.

The human genome consists of all the DNA of our species, the hereditary code of life. This newly revealed text was 3 billion letters long, and written in a strange and cryptographic four-letter code. Such is the amazing complexity of the information carried within each cell of the human body, that a live reading of that code at a rate of three letters per second would take thirty-one years, even if reading continued day and night. Printing these letters out in regular font size on normal bond paper and binding them all together would result in a tower the height of the Washington Monument. For the first time on that summer morning this amazing script, carrying within it all of the instructions for building a human being, was available to the world.

As the leader of the international Human Genome Project, which had labored mightily over more than a decade to reveal this DNA sequence, I stood beside President Bill Clinton in the East Room of the White House, along with Craig Venter, the leader of a competing private sector enterprise. Prime Minister Tony Blair was connected to the event by satellite, and celebrations were occurring simultaneously in many parts of the world.

Clinton's speech began by comparing this human sequence map to the map that Meriwether Lewis had unfolded in front of President Thomas Jefferson in that very room nearly two hundred years earlier. Clinton said, "Without a doubt, this is the most important, most wondrous map ever produced by humankind." But the part of his speech that most attracted public attention jumped from the scientific perspective to the spiritual. "Today," he said, "we are learning the language in which God created life. We are gaining ever more awe for the complexity, the beauty, and the wonder of God's most divine and sacred gift."

Was I, a rigorously trained scientist, taken aback at such a blatantly religious reference by the leader of the free world at a moment such as this? Was I tempted to scowl or look at the floor in embarrassment? No, not at all. In fact I had worked closely with the president's speechwriter in the frantic days just prior to this announcement, and had strongly endorsed the inclusion of this paragraph. When it came time for me to add a few words of my own, I echoed this sentiment: "It's a happy day for the world. It is humbling for me, and awe-inspiring, to realize that we have caught the first glimpse of our own instruction book, previously known only to God."

What was going on here? Why would a president and a scientist, charged with announcing a milestone in biology and medicine, feel compelled to invoke a connection with God? Aren't the scientific and spiritual worldviews antithetical, or shouldn't they at least avoid appearing in the East Room together? What were the reasons for invoking God in these two speeches? Was this poetry? Hypocrisy? A cynical attempt to curry favor from believers, or to disarm those who might criticize this study of the human genome as reducing humankind to machinery? No. Not for me. Quite the contrary, for me the experience of sequencing the human genome, and uncovering this most remarkable of all texts, was both a stunning scientific achievement and an occasion of worship.

Many will be puzzled by these sentiments, assuming that a rigorous scientist could not also be a serious believer in a transcendent God. This book aims to dispel that notion, by arguing that belief in God can be an entirely rational choice, and that the principles of faith are, in fact, complementary with the principles of science.

This potential synthesis of the scientific and spiritual worldviews is assumed by many in modern times to be an impossibility, rather like trying to force the two poles of a magnet together into the same spot. Despite that impression, however, many Americans seem interested in incorporating the validity of both of these worldviews into their daily lives. Recent polls confirm that 93 percent of Americans profess some form of belief in God; yet most of them also drive cars, use electricity, and pay attention to weather reports, apparently assuming that the science undergirding these phenomena is generally trustworthy.

And what about spiritual belief amongst scientists? This is actually more prevalent than many realize. In 1916, researchers asked biologists, physicists, and mathematicians whether they believed in a God who actively communicates with humankind and to whom one may pray in expectation of receiving an answer. About 40 percent answered in the affirmative. In 1997, the same survey was repeated verbatim—and to the surprise of the researchers, the percentage remained very nearly the same.

So perhaps the "battle" between science and religion is not as polarized as it seems? Unfortunately, the evidence of potential harmony is often overshadowed by the high-decibel pronouncements of those who occupy the poles of the debate. Bombs are definitely being thrown from both sides. For example, essentially discrediting the spiritual beliefs of 40 percent of his colleagues as sentimental nonsense, the prominent evolutionist Richard Dawkins has emerged as the leading spokesperson for the point of view that a belief in evolution demands atheism. Among his many eye-popping statements: "Faith is the great cop-out, the great excuse to evade the need to think and evaluate evidence. Faith is belief in spite of, even perhaps because of, the lack of evidence. . . . Faith, being belief that isn't based on evidence, is the principal vice of any religion."[1]

On the other side, certain religious fundamentalists attack science as dangerous and untrustworthy, and point to a literal interpretation of sacred texts as the only reliable means of discerning scientific truth. Among this community, comments from the late Henry Morris, a leader of the creationist movement, stand out: "Evolution's lie permeates and dominates modern thought in every field. That being the case, it follows inevitably that evolutionary thought is basically responsible for the lethally ominous political developments, and the chaotic moral and social disintegrations that have been accelerating everywhere. . . . When science and the Bible differ, science has obviously misinterpreted its data."[2]

This rising cacophony of antagonistic voices leaves many sincere observers confused and disheartened. Reasonable people conclude that they are forced to choose between these two unappetizing extremes, neither of which offers much comfort. Disillusioned by the stridency of both perspectives, many choose to reject both the trustworthiness of scientific conclusions and the value of organized religion, slipping instead into various forms of antiscientific thinking, shallow spirituality, or simple apathy. Others decide to accept the value of both science and spirit, but compartmentalize these parts of their spiritual and material existence to avoid any uneasiness about apparent conflicts. Along these lines, the late biologist Stephen Jay Gould advocated that science and faith should occupy separate, "non-overlapping magisteria." But this, too, is potentially unsatisfying. It inspires internal conflict, and deprives people of the chance to embrace either science or spirit in a fully realized way.

So here is the central question of this book: In this modern era of cosmology, evolution, and the human genome, is there still the possibility of a richly satisfying harmony between the scientific and spiritual worldviews? I answer with a resounding *yes*! In my view, there is no conflict in being a rigorous scientist and a person who believes in a God who takes a personal interest in each one of us. Science's domain is to explore nature. God's domain is in the spiritual world, a realm not possible to explore with the tools and language of science. It must be examined with the heart, the mind, and the soul—and the mind must find a way to embrace both realms.

I will argue that these perspectives not only can coexist within one person, but can do so in a fashion that enriches and enlightens the human experience. Science is the only reliable way to understand the natural world, and its tools when properly utilized can generate profound insights into material existence. But science is powerless to answer questions such as "Why did the universe come into being?" "What is the meaning of human existence?" "What happens after we die?" One of the strongest motivations of humankind is to seek answers to profound questions, and we need to bring all the power of both the scientific and spiritual perspectives to bear on understanding what is both seen and unseen. The goal of this book is to explore a pathway toward a sober and intellectually honest integration of these views.

The consideration of such weighty matters can be unsettling. Whether we call it by name or not, all of us have arrived at a certain worldview. It helps us make sense of the world around us, provides us with an ethical framework, and guides our decisions about the future. Anyone who tinkers with that worldview should not do it lightly. A book that proposes to challenge something so fundamental may inspire more uneasiness than comfort. But we humans seem to possess a deepseated longing to find the truth, even though that longing is easily suppressed by the mundane details of daily life. Those distractions combine with a desire to avoid considering our own mortality, so that days, weeks, months, or even years can easily pass where no serious consideration is given to the eternal questions of human existence. This book is only a small antidote to that circumstance, but will perhaps provide an opportunity for self-reflection, and a desire to look deeper.

First, I should explain how a scientist who studies genetics came to be a believer in a God who is unlimited by time and space, and who takes personal interest in human beings. Some will assume that this must have come about by rigorous religious upbringing, deeply instilled by family and culture, and thus inescapable in later life. But that's not really my story.

PART ONE

The Chasm Between Science and Faith

CHAPTER ONE

From Atheism to Belief

Y EARLY LIFE WAS UNCONVENTIONAL in many ways, but as the son of freethinkers, I had an upbringing that was quite conventionally modern in its attitude toward faith—it just wasn't very important.

I was raised on a dirt farm in the Shenandoah Valley of Virginia. The farm had no running water, and few other physical amenities. Yet these things were more than compensated for by the stimulating mix of experiences and opportunities that were available to me in the remarkable culture of ideas created by my parents.

They had met in graduate school at Yale in 1931, and had taken their community organizing skills and love of music to the experimental community of Arthurdale, West Virginia, where they worked with Eleanor Roosevelt in attempting to reinvigorate a downtrodden mining community in the depths of the Great Depression.

But other advisers in the Roosevelt administration had other ideas, and the funding soon dried up. The ultimate dismantling of the Arthurdale community on the basis of backbiting Washington politics left my parents with a lifelong suspicion of the government. They moved on to academic life at Elon College in Burlington, North Carolina. There, presented with the wild and beautiful folk culture of the rural South, my father became a folksong collector, traveling through the hills and hollows and convincing reticent North Carolinians to sing into his Presto recorder. Those recordings, along with an even larger set from Alan Lomax, make up a significant fraction of the Library of Congress collection of American folksongs.

When World War II arrived, such musical endeavors were forced to take a backseat to more urgent matters of national defense, and my father went to work helping to build bombers for the war effort, ultimately ending up as a supervisor in an aircraft factory in Long Island.

At the end of the war, my parents concluded that the high-pressure life of business was not for them. Being ahead of their time, they did the "sixties thing" in the 1940s: they moved to the Shenandoah Valley of Virginia, bought a ninety-five-acre farm, and set about trying to create a simple agricultural lifestyle without use of farm machinery. Discovering after only a few months that this was not going to feed their two adolescent sons (and soon another brother and I would arrive), my father landed a job teaching drama at the local women's college. He recruited male actors from the local town, and together these college students and local tradesmen found the production of plays was great fun. Faced with complaints about the long and boring hiatus in the summer, my father and mother founded a summer theater in a grove of oak trees above our farmhouse. The Oak Grove Theater continues in uninterrupted and delightful operation more than fifty years later.

I was born into this happy mix of pastoral beauty, hard farmwork, summer theater, and music, and thrived in it. As the youngest of four boys, I could not get into too many scrapes that were not already familiar to my parents. I grew up with the general sense that you had to be responsible for your own behavior and your choices, as no one else was going to step in and take care of them for you.

Like my older brothers, I was home-schooled by my mother, a remarkably talented teacher. Those early years conferred on me the priceless gift of the joy of learning. While my mother had no organized class schedule or lesson plans, she was incredibly perceptive in identifying topics that would intrigue a young mind, pursuing them with great intensity to a natural stopping point, and then switching to something new and equally exciting. Learning was never something you did because you had to, it was something you did because you loved it.

Faith was not an important part of my childhood. I was vaguely aware of the concept of God, but my own interactions with Him were limited to occasional childish moments of bargaining about something that I really wanted Him to do for me. For instance, I remember making a contract with God (at about age nine) that if He would prevent the rainout of a Saturday night theater performance and music party that I was particularly excited about, then I would promise never to smoke cigarettes. Sure enough, the rains held off, and I never took up the habit. Earlier, when I was five, my parents decided to send me and my next oldest brother to become members of the boys choir at the local Episcopal church. They made it clear that it would be a great way to learn music, but that the theology should not be taken too seriously. I followed those instructions, learning the glories of harmony and counterpoint but letting the theological concepts being preached from the pulpit wash over me without leaving any discernible residue.

When I was ten, we moved into town to be with my ailing grandmother, and I entered the public schools. At fourteen, my eyes were opened to the wonderfully exciting and powerful methods of science. Inspired by a charismatic chemistry teacher who could write the same information on the blackboard with both hands simultaneously, I discovered for the first time the intense satisfaction of the ordered nature of the universe. The fact that all matter was constructed of atoms and molecules that followed mathematical principles was an unexpected revelation, and the ability to use the tools of science to discover new things about nature struck me at once as something of which I wanted to be a part. With the enthusiasm of a new convert, I decided my goal in life would be to become a chemist. Never mind that I knew relatively little about the other sciences, this first puppy love seemed life-changing.

In contrast, my encounters with biology left me completely cold. At least as perceived by my teenage mind, the fundamentals of biology seemed to have more to do with rote learning of mindless facts than elucidation of principles. I really wasn't that interested in memorizing the parts of the crayfish, nor in trying to figure out the difference between a phylum, a class, and an order. The overwhelming complexity of life led me to the conclusion that biology was rather like existential philosophy: it just didn't make sense. For my budding reductionist mind, there was not nearly enough logic in it to be appealing. Graduating at sixteen, I went on to the University of Virginia, determined to major in chemistry and pursue a scientific career. Like most college freshmen, I found this new environment invigorating, with so many ideas bouncing off the classroom walls and in the dorm rooms late at night. Some of those questions invariably turned to the existence of God. In my early teens I had had occasional moments of the experience of longing for something outside myself, often associated with the beauty of nature or a particularly profound musical experience. Nevertheless, my sense of the spiritual was very undeveloped and easily challenged by the one or two aggressive atheists one finds in almost every college dormitory. By a few months into my college career, I became convinced that while many religious faiths had inspired interesting traditions of art and culture, they held no foundational truth.

THOUGH I DID NOT KNOW the term at the time, I became an agnostic, a term coined by the nineteenth-century scientist T. H. Huxley to indicate someone who simply does not know whether or not God exists. There are all kinds of agnostics;

some arrive at this position after intense analysis of the evidence, but many others simply find it to be a comfortable position that allows them to avoid considering arguments they find discomforting on either side. I was definitely in the latter category. In fact, my assertion of "I don't know" was really more along the lines of "I don't want to know." As a young man growing up in a world full of temptations, it was convenient to ignore the need to be answerable to any higher spiritual authority. I practiced a thought and behavior pattern referred to as "willful blindness" by the noted scholar and writer C. S. Lewis.

After graduation, I went on to a Ph.D. program in physical chemistry at Yale, pursuing the mathematical elegance that had first drawn me to this branch of science. My intellectual life was immersed in quantum mechanics and second-order differential equations, and my heroes were the giants of physics—Albert Einstein, Niels Bohr, Werner Heisenberg, and Paul Dirac. I gradually became convinced that everything in the universe could be explained on the basis of equations and physical principles. Reading the biography of Albert Einstein, and discovering that despite his strong Zionist position after World War II, he did not believe in Yahweh, the God of the Jewish people, only reinforced my conclusion that no thinking scientist could seriously entertain the possibility of God without committing some sort of intellectual suicide.

And so I gradually shifted from agnosticism to atheism. I felt quite comfortable challenging the spiritual beliefs of anyone who mentioned them in my presence, and discounted such perspectives as sentimentality and outmoded superstition.

Two years into this Ph.D. program my narrowly structured

life plan began to come apart. Despite the daily pleasures of pursuing my dissertation research on theoretical quantum mechanics, I began to doubt whether this would be a life-sustaining pathway for me. It seemed that most of the major advances in quantum theory had occurred fifty years earlier, and most of my career was likely to be spent in applying successive simplifications and approximations to render certain elegant but unsolvable equations just a tiny bit more tractable. More practically, it seemed that my path would lead inexorably to a professor's life of delivering an interminable series of lectures on thermodynamics and statistical mechanics, presented to class after class of undergraduates who were either bored or terrified by those subjects.

At about that same time, in an effort to broaden my horizons, I signed up for a course in biochemistry, finally investigating the life sciences that I had so carefully avoided in the past. The course was nothing short of astounding. The principles of DNA, RNA, and protein, never previously apparent to me, were laid out in all of their satisfying digital glory. The ability to apply rigorous intellectual principles to understanding biology, something I had assumed impossible, was bursting forth with the revelation of the genetic code. With the advent of new methods for splicing different DNA fragments together at will (recombinant DNA), the possibility of applying all of this knowledge for human benefit seemed quite real. I was astounded. Biology has mathematical elegance after all. Life makes sense.

At the same time, now only twenty-two but married with a bright and inquisitive daughter, I was becoming more social. I had often preferred to be alone when I was younger. Now, human interaction and a desire to contribute something to humanity seemed ever more important. Putting all of these sudden revelations together, I questioned everything about my previous choices, including whether I was really cut out to do science or carry out independent research. I was just about to complete my Ph.D., yet after much soul-searching, I applied for admission to medical school. With a carefully practiced speech, I attempted to convince admissions committees that this turn of events was actually a natural pathway for the training of one of our nation's future doctors. Inside I was not so sure. After all, wasn't I the guy who had hated biology because you had to memorize things? Could any field of study require more memorization than medicine? But something was different now: this was about humanity, not crayfish; there were principles underlying the details; and this could ultimately make a difference in the lives of real people.

I was accepted at the University of North Carolina. Within a few weeks I knew medical school was the right place for me. I loved the intellectual stimulation, the ethical challenges, the human element, and the amazing complexity of the human body. In December of that first year I found out how to combine this new love of medicine with my old love of mathematics. An austere and somewhat unapproachable pediatrician, who taught a grand total of six hours of lectures on medical genetics to the first-year medical student class, showed me my future. He brought patients to class with sickle cell anemia, galactosemia (an often-fatal inability to tolerate milk products), and Down syndrome, all caused by glitches in the genome, some as subtle as a single letter gone awry.

I was astounded by the elegance of the human DNA code, and the multiple consequences of those rare careless moments of its copying mechanism. Though the potential to actually do anything to help very many of those afflicted by such genetic diseases seemed far away, I was immediately drawn to this discipline. While at that point no shadow of possibility of anything as grand and consequential as the Human Genome Project had entered a single human mind, the path I started on in December of 1973 turned out fortuitously to lead directly into participation in one of the most historic undertakings of humankind.

This path also led me by the third year of medical school into intense experiences involving the care of patients. As physicians in training, medical students are thrust into some of the most intimate relationships imaginable with individuals who had been complete strangers until their experience of illness. Cultural taboos that normally prevent the exchange of intensely private information come tumbling down along with the sensitive physical contact of a doctor and his patients. It is all part of the long-standing and venerated contract between the ill person and the healer. I found the relationships that developed with sick and dying patients almost overwhelming, and I struggled to maintain the professional distance and lack of emotional involvement that many of my teachers advocated.

What struck me profoundly about my bedside conversations with these good North Carolina people was the spiritual aspect of what many of them were going through. I witnessed numerous cases of individuals whose faith provided them with a strong reassurance of ultimate peace, be it in this world or the next, despite terrible suffering that in most instances they had done nothing to bring on themselves. If faith was a psychological crutch, I concluded, it must be a very powerful one. If it was nothing more than a veneer of cultural tradition, why were these people not shaking their fists at God and demanding that their friends and family stop all this talk about a loving and benevolent supernatural power?

My most awkward moment came when an older woman, suffering daily from severe untreatable angina, asked me what I believed. It was a fair question; we had discussed many other important issues of life and death, and she had shared her own strong Christian beliefs with me. I felt my face flush as I stammered out the words "I'm not really sure." Her obvious surprise brought into sharp relief a predicament that I had been running away from for nearly all of my twenty-six years: I had never really seriously considered the evidence for and against belief.

That moment haunted me for several days. Did I not consider myself a scientist? Does a scientist draw conclusions without considering the data? Could there be a more important question in all of human existence than "Is there a God?" And yet there I found myself, with a combination of willful blindness and something that could only be properly described as arrogance, having avoided any serious consideration that God might be a real possibility. Suddenly all my arguments seemed very thin, and I had the sensation that the ice under my feet was cracking.

This realization was a thoroughly terrifying experience. After all, if I could no longer rely on the robustness of my atheistic position, would I have to take responsibility for actions that I would prefer to keep unscrutinized? Was I answerable to someone other than myself? The question was now too pressing to avoid.

At first, I was confident that a full investigation of the rational basis for faith would deny the merits of belief, and reaffirm my atheism. But I determined to have a look at the facts, no matter what the outcome. Thus began a quick and confusing survey through the major religions of the world. Much of what I found in the CliffsNotes versions of different religions (I found reading the actual sacred texts much too difficult) left me thoroughly mystified, and I found little reason to be drawn to one or the other of the many possibilities. I doubted that there was any rational basis for spiritual belief undergirding any of these faiths. However, that soon changed. I went to visit a Methodist minister who lived down the street to ask him whether faith made any logical sense. He listened patiently to my confused (and probably blasphemous) ramblings, and then took a small book off his shelf and suggested I read it.

The book was *Mere Christianity* by C. S. Lewis. In the next few days, as I turned its pages, struggling to absorb the breadth and depth of the intellectual arguments laid down by this legendary Oxford scholar, I realized that all of my own constructs against the plausibility of faith were those of a schoolboy. Clearly I would need to start with a clean slate to consider this most important of all human questions. Lewis seemed to know all of my objections, sometimes even before I had quite formulated them. He invariably addressed them within a page or two. When I learned subsequently that Lewis had himself been an atheist, who had set out to disprove faith on the basis of logical argument, I recognized how he could be so insightful about my path. It had been his path as well.

The argument that most caught my attention, and most

rocked my ideas about science and spirit down to their foundation, was right there in the title of Book One: "Right and Wrong as a Clue to the Meaning of the Universe." While in many ways the "Moral Law" that Lewis described was a universal feature of human existence, in other ways it was as if I was recognizing it for the first time.

To understand the Moral Law, it is useful to consider, as Lewis did, how it is invoked in hundreds of ways each day without the invoker stopping to point out the foundation of his argument. Disagreements are part of daily life. Some are mundane, as the wife criticizing her husband for not speaking more kindly to a friend, or a child complaining, "It's not fair," when different amounts of ice cream are doled out at a birthday party. Other arguments take on larger significance. In international affairs, for instance, some argue that the United States has a moral obligation to spread democracy throughout the world, even if it requires military force, whereas others say that the aggressive, unilateral use of military and economic force threatens to squander moral authority.

In the area of medicine, furious debates currently surround the question of whether or not it is acceptable to carry out research on human embryonic stem cells. Some argue that such research violates the sanctity of human life; others posit that the potential to alleviate human suffering constitutes an ethical mandate to proceed. (This topic and several other dilemmas in bioethics are considered in the Appendix to this book.)

Notice that in all these examples, each party attempts to appeal to an unstated higher standard. This standard is the Moral Law. It might also be called "the law of right behavior," and its existence in each of these situations seems unquestioned. What is being debated is whether one action or another is a closer approximation to the demands of that law. Those accused of having fallen short, such as the husband who is insufficiently cordial to his wife's friend, usually respond with a variety of excuses why they should be let off the hook. Virtually never does the respondent say, "To hell with your concept of right behavior."

What we have here is very peculiar: the concept of right and wrong appears to be universal among all members of the human species (though its application may result in wildly different outcomes). It thus seems to be a phenomenon approaching that of a law, like the law of gravitation or of special relativity. Yet in this instance, it is a law that, if we are honest with ourselves, is broken with astounding regularity.

As best as I can tell, this law appears to apply peculiarly to human beings. Though other animals may at times appear to show glimmerings of a moral sense, they are certainly not widespread, and in many instances other species' behavior seems to be in dramatic contrast to any sense of universal rightness. It is the awareness of right and wrong, along with the development of language, awareness of self, and the ability to imagine the future, to which scientists generally refer when trying to enumerate the special qualities of *Homo sapiens*.

But is this sense of right and wrong an intrinsic quality of being human, or just a consequence of cultural traditions? Some have argued that cultures have such widely differing norms for behavior that any conclusion about a shared Moral Law is unfounded. Lewis, a student of many cultures, calls this "a lie, a good resounding lie. If a man will go into a library and spend a few days with the *Encyclopedia of Religion and Ethics,* he will soon discover the massive unanimity of the practical reason in man. From the Babylonian Hymn to Samos, from the laws of Manu, the Book of the Dead, the Analects, the Stoics, the Platonists, from Australian aborigines and Redskins, he will collect the same triumphantly monotonous denunciations of oppression, murder, treachery and falsehood; the same injunctions of kindness to the aged, the young, and the weak, of almsgiving and impartiality and honesty."[1] In some unusual cultures the law takes on surprising trappings—consider the execution of suspected witches in seventeenth-century America. Yet when surveyed closely, these apparent aberrations can be seen to arise from strongly held but misguided conclusions about who or what is good or evil. If you firmly believed that a witch is the personification of evil on earth, an apostle of the devil himself, would it not then seem justified to take such drastic action?

Let me stop here to point out that the conclusion that the Moral Law exists is in serious conflict with the current postmodernist philosophy, which argues that there are no absolute rights or wrongs, and all ethical decisions are relative. This view, which seems widespread among modern philosophers but which mystifies most members of the general public, faces a series of logical Catch-22s. If there is no absolute truth, can postmodernism itself be true? Indeed, if there is no right or wrong, then there is no reason to argue for the discipline of ethics in the first place.

Others will object that the Moral Law is simply a consequence of evolutionary pressures. This objection arises from the new field of sociobiology, and attempts to provide explanations for altruistic behavior on the basis of its positive value in Darwinian selection. If this argument could be shown to hold up, the interpretation of many of the requirements of the Moral Law as a signpost to God would potentially be in trouble—so it is worth examining this point of view in more detail.

Consider a major example of the force we feel from the Moral Law—the altruistic impulse, the voice of conscience calling us to help others even if nothing is received in return. Not all of the requirements of the Moral Law reduce to altruism, of course; for instance, the pang of conscience one feels after a minor distortion of the facts on a tax return can hardly be ascribed to a sense of having damaged another identifiable human being.

First, let's be clear what we're talking about. By altruism I do not mean the "You scratch my back, I'll scratch yours" kind of behavior that practices benevolence to others in direct expectation of reciprocal benefits. Altruism is more interesting: the truly selfless giving of oneself to others with absolutely no secondary motives. When we see that kind of love and generosity, we are overcome with awe and reverence. Oskar Schindler placed his life in great danger by sheltering more than a thousand Jews from Nazi extermination during World War II, and ultimately died penniless—and we feel a great rush of admiration for his actions. Mother Teresa has consistently ranked as one of the most admired individuals of the current age, though her self-imposed poverty and selfless giving to the sick and dying of Calcutta is in drastic contrast to the materialistic lifestyle that dominates our current culture.

In some instances, altruism can extend even to circumstances where the beneficiary would seem to be a sworn enemy. Sister Joan Chittister, a Benedictine nun, tells the following Sufi story.[2]

> Once upon a time there was an old woman who used to meditate on the bank of the Ganges. One morning, finishing her meditation, she saw a scorpion floating helplessly in the strong current. As the scorpion was pulled closer, it got caught in roots that branched out far into the river. The scorpion struggled frantically to free itself but got more and more entangled. She immediately reached out to the drowning scorpion, which, as soon as she touched it, stung her. The old woman withdrew her hand but, having regained her balance, once again tried to save the creature. Every time she tried, however, the scorpion's tail stung her so badly that her hands became bloody and her face distorted with pain. A passerby who saw the old woman struggling with the scorpion shouted, "What's wrong with you, fool! Do you want to kill yourself to save that ugly thing?" Looking into the stranger's eyes, she answered, "Because it is the nature of the scorpion to sting, why should I deny my own nature to save it?"

This may seem a rather drastic example—not very many of us can relate to putting ourselves in danger to save a scorpion.

But surely most of us have at one time felt the inner calling to help a stranger in need, even with no likelihood of personal benefit. And if we have actually acted on that impulse, the consequence was often a warm sense of "having done the right thing."

C. S. Lewis, in his remarkable book *The Four Loves,* further explores the nature of this kind of selfless love, which he calls "agape" (pronounced *ah-GAH-pay*), from the Greek. He points out that this kind of love can be distinguished from the three other forms (affection, friendship, and romantic love), which can be more easily understood in terms of reciprocal benefit, and which we can see modeled in other animals besides ourselves.

Agape, or selfless altruism, presents a major challenge for the evolutionist. It is quite frankly a scandal to reductionist reasoning. It cannot be accounted for by the drive of individual selfish genes to perpetuate themselves. Quite the contrary: it may lead humans to make sacrifices that lead to great personal suffering, injury, or death, without any evidence of benefit. And yet, if we carefully examine that inner voice we sometimes call conscience, the motivation to practice this kind of love exists within all of us, despite our frequent efforts to ignore it.

Sociobiologists such as E. O. Wilson have attempted to explain this behavior in terms of some indirect reproductive benefits to the practitioner of altruism, but the arguments quickly run into trouble. One proposal is that repeated altruistic behavior of the individual is recognized as a positive attribute in mate selection. But this hypothesis is in direct conflict with observations in nonhuman primates that often reveal just the opposite—such as the practice of infanticide by a newly dominant male monkey, in order to clear the way for his own future offspring. Another argument is that there are indirect reciprocal benefits from altruism that have provided advantages to the practitioner over evolutionary time; but this explanation cannot account for human motivation to practice small acts of conscience that no one else knows about. A third argument is that altruistic behavior by members of a group provides benefits to the whole group. Examples are offered of ant colonies, where sterile workers toil incessantly to create an environment where their mothers can have more children. But this kind of "ant altruism" is readily explained in evolutionary terms by the fact that the genes motivating the sterile worker ants are *exactly* the same ones that will be passed on by their mother to the siblings they are helping to create. That unusually direct DNA connection does not apply to more complex populations, where evolutionists now agree almost universally that selection operates on the individual, not on the population. The hardwired behavior of the worker ant is thus fundamentally different from the inner voice that causes me to feel compelled to jump into the river to try to save a drowning stranger, even if I'm not a good swimmer and may myself die in the effort. Furthermore, for the evolutionary argument about group benefits of altruism to hold, it would seem to require an opposite response, namely, hostility to individuals outside the group. Oskar Schindler's and Mother Teresa's agape belies this kind of thinking. Shockingly, the Moral Law will ask me to save the drowning man even if he is an enemy.

If the Law of Human Nature cannot be explained away as

cultural artifact or evolutionary by-product, then how can we account for its presence? There is truly something unusual going on here. To quote Lewis, "If there was a controlling power outside the universe, it could not show itself to us as one of the facts inside the universe—no more than the architect of a house could actually be a wall or staircase or fireplace in that house. The only way in which we could expect it to show itself would be inside ourselves as an influence or a command trying to get us to behave in a certain way. And that is just what we do find inside ourselves. Surely this ought to arouse our suspicions?"[3]

Encountering this argument at age twenty-six, I was stunned by its logic. Here, hiding in my own heart as familiar as anything in daily experience, but now emerging for the first time as a clarifying principle, this Moral Law shone its bright white light into the recesses of my childish atheism, and demanded a serious consideration of its origin. Was this God looking back at me?

And if that were so, what kind of God would this be? Would this be a deist God, who invented physics and mathematics and started the universe in motion about 14 billion years ago, then wandered off to deal with other, more important matters, as Einstein thought? No, this God, if I was perceiving Him at all, must be a theist God, who desires some kind of relationship with those special creatures called human beings, and has therefore instilled this special glimpse of Himself into each one of us. This might be the God of Abraham, but it was certainly not the God of Einstein.

There was another consequence to this growing sense of

God's nature, if in fact He was real. Judging by the incredibly high standards of the Moral Law, one that I had to admit I was in the practice of regularly violating, this was a God who was holy and righteous. He would have to be the embodiment of goodness. He would have to hate evil. And there was no reason to suspect that this God would be kindly or indulgent. The gradual dawning of my realization of God's plausible existence brought conflicted feelings: comfort at the breadth and depth of the existence of such a Mind, and yet profound dismay at the realization of my own imperfections when viewed in His light.

I had started this journey of intellectual exploration to confirm my atheism. That now lay in ruins as the argument from the Moral Law (and many other issues) forced me to admit the plausibility of the God hypothesis. Agnosticism, which had seemed like a safe second-place haven, now loomed like the great cop-out it often is. Faith in God now seemed more rational than disbelief.

It also became clear to me that science, despite its unquestioned powers in unraveling the mysteries of the natural world, would get me no further in resolving the question of God. If God exists, then He must be outside the natural world, and therefore the tools of science are not the right ones to learn about Him. Instead, as I was beginning to understand from looking into my own heart, the evidence of God's existence would have to come from other directions, and the ultimate decision would be based on faith, not proof. Still beset by roiling uncertainties of what path I had started down, I had to admit that I had reached the threshold of accepting the possibility of a spiritual worldview, including the existence of God.

It seemed impossible either to go forward or to turn back. Years later, I encountered a sonnet by Sheldon Vanauken that precisely described my dilemma. Its concluding lines:

> Between the probable and proved there yawns
> A gap. Afraid to jump, we stand absurd,
> Then see *behind* us sink the ground and, worse,
> Our very standpoint crumbling. Desperate dawns
> Our only hope: to leap into the Word
> That opens up the shuttered universe.[4]

For a long time I stood trembling on the edge of this yawning gap. Finally, seeing no escape, I leapt.

How can such beliefs be possible for a scientist? Aren't many claims of religion incompatible with the "Show me the data" attitude of someone devoted to the study of chemistry, physics, biology, and medicine? By opening the door of my mind to its spiritual possibilities, had I started a war of worldviews that would consume me, ultimately facing a take-no-prisoners victory of one or the other?

CHAPTER TWO

The War of the Worldviews

I F YOU STARTED THIS BOOK as a skeptic and have managed to travel this far with me, no doubt a torrent of your own objections has begun to form. I certainly have had my own: Isn't God just a case of wishful thinking? Hasn't a great deal of harm been done in the name of religion? How could a loving God permit suffering? How can a serious scientist accept the possibility of miracles?

If you are a believer, perhaps the narrative in the first chapter offered some reassurance, but almost certainly you, too, have areas where your faith conflicts with other challenges you face from yourself or those around you.

Doubt is an unavoidable part of belief. In the words of Paul Tillich, "Doubt isn't the opposite of faith; it is an element of faith."[1] If the case in favor of belief in God were utterly airtight, then the world would be full of confident practitioners of a single faith. But imagine such a world, where the opportunity to make a free choice about belief was taken away by the certainty of the evidence. How interesting would that be?

For the skeptic and the believer alike, doubts come from many sources. One category involves perceived conflicts of the claims of religious belief with scientific observations. Those concerns, particularly prominent now in the field of biology and genetics, are dealt with in subsequent chapters. Other concerns reside more within the philosophical realm of human experience, and those are the subject of this chapter. If you are not someone who is troubled by these, then feel free to turn to Chapter 3.

In addressing these philosophical issues, I speak mainly as a layman. Yet I am one who has shared these struggles. Especially in the first year after I came to accept the existence of a God who cared about human beings, I was besieged by doubts from many directions. While these questions all seemed very fresh and unanswerable upon their first arrival, I was comforted to learn that there were no objections on my list that had not been raised even more forcefully and articulately by others down through the centuries. Of greatest comfort, many wonderful sources existed that provided compelling answers to these dilemmas. I will draw upon some of these authors in this chapter, supplemented by my own thoughts and experiences. Many of the most accessible analyses came from the writings of my now familiar Oxford adviser, C. S. Lewis.

While many objections could be considered here, I found four to be particularly vexing in those early days of newborn faith, and I believe these are among the top concerns faced by anyone considering a decision about belief in God.

Isn't the Idea of God Just Wish Fulfillment?

Is God really there? Or does the search for the existence of a supernatural being, so pervasive in all cultures ever studied, represent a universal but groundless human longing for something outside ourselves to give meaning to a meaningless life and to take away the sting of death?

While the search for the divine has been somewhat crowded out in modern times by our busy and overstimulated lives, it is still one of the most universal of human strivings. C. S. Lewis describes this phenomenon in his own life in his wonderful book *Surprised by Joy*, and it is this sense of intense longing, triggered in his life by something as simple as a few lines of poetry, that he identifies as "joy." He describes the experience as "an unsatisfied desire which is itself more desirable than any other satisfaction."[2] I can recall clearly some of those moments in my own life, where this poignant sense of longing, falling somewhere between pleasure and grief, caught me by surprise and caused me to wonder from whence came such strong emotion, and how might such an experience be recovered.

As a boy of ten, I recall being transported by the experience of looking through a telescope that an amateur astronomer had placed on a high field at our farm, when I sensed the vastness of the universe and saw the craters on the moon and the magical diaphanous light of the Pleiades. At fifteen, I recall a Christmas Eve where the descant on a particularly beautiful Christmas carol, rising sweet and true above the more familiar tune, left me with a sense of unexpected awe and a longing for something I could not name. Much later, as an atheist graduate student, I surprised myself by experiencing this same sense of awe and longing, this time mixed with a particularly deep sense of grief, at the playing of the second movement of Beethoven's Third Symphony (the *Eroica*). As the world grieved the death of Israeli athletes killed by terrorists at the Olympics in 1972, the Berlin Philharmonic played the powerful strains of this C-minor lament in the Olympic Stadium, mixing together nobility and tragedy, life and death. For a few moments I was lifted out of my materialist worldview into an indescribable spiritual dimension, an experience I found quite astonishing.

More recently, for a scientist who occasionally is given the remarkable privilege of discovering something not previously known by man, there is a special kind of joy associated with such flashes of insight. Having perceived a glimmer of scientific truth, I find at once both a sense of satisfaction and a longing to understand some even greater Truth. In such a moment, science becomes more than a process of discovery. It transports the scientist into an experience that defies a completely naturalistic explanation.

So what are we to make of these experiences? And what is this sensation of longing for something greater than ourselves? Is this only, and no more than, some combination of neurotransmitters landing on precisely the right receptors, setting off an electrical discharge deep in some part of the brain? Or is this, like the Moral Law described in the preceding chapter, an inkling of what lies beyond, a signpost placed deep within the human spirit pointing toward something much grander than ourselves?

The atheist view is that such longings are not to be trusted as indications of the supernatural, and that our translation of those sensations of awe into a belief in God represent nothing more than wishful thinking, inventing an answer because we want it to be true. This particular view reached its widest audience in the writings of Sigmund Freud, who argued that wishes for God stemmed from early childhood experiences. Writing in *Totem and Taboo,* Freud said, "Psychoanalysis of individual human beings teaches us with quite special insistence that the God of each of them is formed in the likeness of his father, that his personal relationship to God depends on the relation to his father in the flesh, and oscillates and changes along with that relation, and that at bottom God is nothing other than an exalted father."[3]

The problem with this wish-fulfillment argument is that it does not accord with the character of the God of the major religions of the earth. In his elegant recent book, *The Question of God,* Armand Nicholi, a psychoanalytically trained Harvard professor, compares Freud's view with that of C. S. Lewis.[4] Lewis argued that such wish fulfillment would likely give rise to a very different kind of God than the one described in the Bible. If we are looking for benevolent coddling and indulgence, that's not what we find there. Instead, as we begin to come to grips with the existence of the Moral Law, and our obvious inability to live up to it, we realize that we are in deep trouble, and are potentially eternally separated from the Author of that Law. Furthermore, does not a child as he or she grows up experience ambivalent feelings toward parents, including a desire to be free? So why should wish fulfillment lead to a desire for God, as opposed to a desire for there to be no God?

Finally, in simple logical terms, if one allows the possibility that God is something humans might wish for, does that rule out the possibility that God is real? Absolutely not. The fact that I have wished for a loving wife does not now make her imaginary. The fact that the farmer wished for rain does not make him question the reality of the subsequent downpour.

In fact, one can turn this wishful-thinking argument on its head. Why would such a universal and uniquely human hunger exist, if it were not connected to some opportunity for fulfillment? Again, Lewis says it well: "Creatures are not born with desires unless satisfaction for those desires exists. A baby feels hunger: well, there is such a thing as food. A duckling wants to swim: well, there is such a thing as water. Men feel sexual desire: well, there is such a thing as sex. If I find in myself a desire which no experience in this world can satisfy, the most probable explanation is that I was made for another world."[5]

Could it be that this longing for the sacred, a universal and puzzling aspect of human experience, may not be wish fulfillment but rather a pointer toward something beyond us? Why do we have a "God-shaped vacuum" in our hearts and minds unless it is meant to be filled?

In our modern materialistic world, it is easy to lose sight of that sense of longing. In her wonderful collection of essays *Teaching a Stone to Talk,* Annie Dillard speaks about that growing void:

Now we are no longer primitive. Now the whole world seems not holy. . . . We as a people have moved from pantheism to pan-atheism. . . . It is difficult to undo our own damage and to recall to our presence that which we have asked to leave. It is hard to desecrate a grove and change your mind. We doused the burning bush and cannot rekindle it. We are lighting matches in vain under every green tree. Did the wind used to cry and the hills shout forth praise? Now speech has perished from among the lifeless things of the earth, and living things say very little to very few. . . . And yet it could be that wherever there is motion there is noise, as when a whale breaches and smacks the water, and wherever there is stillness there is the small, still voice, God's speaking from the whirlwind, nature's old song and dance, the show we drove from town. . . . What have we been doing all these centuries but trying to call God back to the mountain, or, failing that, raise a peep out of anything that isn't us? What is the difference between a cathedral and a physics lab? Are they not both saying: Hello?[6]

WHAT ABOUT ALL THE HARM DONE IN THE NAME OF RELIGION?

A major stumbling block for many earnest seekers is the compelling evidence throughout history that terrible things have been done in the name of religion. This applies to virtually all faiths at some point, including those that argue for compassion and nonviolence among their principal tenets. Given such examples of raw abusive power, violence, and hypocrisy, how can anyone subscribe to the tenets of the faith promoted by such perpetrators of evil?

There are two answers to this dilemma. First of all, keep in mind that many wonderful things have also been done in the name of religion. The church (and here I use the term generically, to refer to the organized institutions that promote a particular faith, without regard to which faith is being described) has many times played a critical role in supporting justice and benevolence. As just one example, consider how religious leaders have worked to relieve people from oppression, from Moses' leading the Israelites out of bondage to William Wilberforce's ultimate victory in convincing the English Parliament to oppose the practice of slavery to the Reverend Martin Luther King Jr.'s leading the civil rights movement in the United States, for which he gave his life.

But the second answer brings us back to the Moral Law, and to the fact that all of us as human beings have fallen short of it. The church is made up of fallen people. The pure, clean water of spiritual truth is placed in rusty containers, and the subsequent failings of the church down through the centuries should not be projected onto the faith itself, as if the water had been the problem. It is no wonder that those who assess the truth and appeal of spiritual faith by the behavior of any particular church often find it impossible to imagine themselves joining up. Expressing hostility toward the French Catholic Church at the dawning of the French Revolution, Voltaire wrote, "Is it any wonder that there are atheists in the world, when the church behaves so abominably?"[7]

It is not difficult to identify examples where the church has promoted actions that fly in the face of principles its own faith should have sustained. The Beatitudes spoken by Christ in the Sermon on the Mount were ignored as the Christian church carried out violent Crusades in the Middle Ages and pursued a series of inquisitions afterward. While in the Mecca phase of his life, the prophet Muhammad never used violence in responding to persecutors, Islamic jihads commenced in the Medina phase and extended over centuries, even to pesent-day violent attacks such as that of September 11, 2001, creating the unfortunate impression that Islam is necessarily violent. Even followers of supposedly nonviolent faiths such as Hinduism and Buddhism occasionally engage in violent confrontation, as is currently occurring in Sri Lanka.

And it is not only violence that sullies the truth of religious faith. Frequent examples of gross hypocrisy among religious leaders, made evermore visible by the power of the media, cause many skeptics to conclude that there is no objective truth or goodness to be found in religion.

Perhaps even more insidious and widespread is the emergence in many churches of a spiritually dead, secular faith, which strips out all of the numinous aspects of traditional belief, presenting a version of spiritual life that is all about social events and/or tradition, and nothing about the search for God.

Is it any wonder, then, that some commentators point to religion as a negative force in society, or in the words of Karl Marx, "the opiate of the masses"? But let's be careful here. The great Marxist experiments in the Soviet Union and in Mao's China, aiming to establish societies explicitly based upon atheism, proved capable of committing at least as much, and probably more, human slaughter and raw abuse of power than the worst of all regimes in recent times. In fact, by denying the existence of any higher authority, atheism has the now-realized potential to free humans completely from any responsibility not to oppress one another.

So, while the long history of religious oppression and hypocrisy is profoundly sobering, the earnest seeker must look beyond the behavior of flawed humans in order to find the truth. Would you condemn an oak tree because its timbers had been used to build battering rams? Would you blame the air for allowing lies to be transmitted through it? Would you judge Mozart's *The Magic Flute* on the basis of a poorly rehearsed performance by fifth-graders? If you had never seen a real sunset over the Pacific, would you allow a tourist brochure as a substitute? Would you evaluate the power of romantic love solely in the light of an abusive marriage next door?

No. A real evaluation of the truth of faith depends upon looking at the clean, pure water, not at the rusty containers.

Why Would a Loving God Allow Suffering in the World?

There may be those somewhere in the world who have never experienced suffering. I don't know any such people, and I suspect no reader of this book would claim to be in that category. This universal human experience has caused many to question the existence of a loving God. As phrased by C. S. Lewis in *The Problem of Pain,* the argument goes like this: "If God were good, he would wish to make his creatures perfectly happy, and if God were almighty, he would be able to do what he wished. But the creatures are not happy. Therefore, God lacks either goodness or power or both."[8]

There are several answers to this dilemma. Some are easier to accept than others. In the first place, let us recognize that a large fraction of our suffering and that of our fellow human beings is brought about by what we do to one another. It is humankind, not God, that has invented knives, arrows, guns, bombs, and all manner of other instruments of torture used through the ages. The tragedy of the young child killed by a drunk driver, of the innocent man dying on the battlefield, or of the young girl cut down by a stray bullet in a crime-ridden section of a modern city can hardly be blamed on God. After all, we have somehow been given free will, the ability to do as we please. We use this ability frequently to disobey the Moral Law. And when we do so, we shouldn't then blame God for the consequences.

Should God have restrained our free will in order to prevent these kinds of evil behavior? That line of thought quickly encounters a dilemma from which there is no rational escape. Again, Lewis states this clearly: "If you choose to say 'God can give a creature free will and at the same time withhold free will from it,' you have not succeeded in saying anything about God: meaningless combinations of words do not suddenly acquire meaning simply because we prefix to them the two other words 'God can.' Nonsense remains nonsense, even when we talk it about God."[9]

Rational arguments can still be difficult to accept when an experience of terrible suffering falls on an innocent person. I know a young college student who was living alone during summer vacation while she carried out medical research in preparation for a career as a physician. Awakening in the dark of night, she found a strange man had broken into her apartment. With a knife pressed against her throat, he ignored her pleas, blindfolded her, and forced himself on her. He left her in devastation, to relive that experience over and over again for years to come. The perpetrator was never caught.

That young woman was my daughter. Never was pure evil more apparent to me than that night, and never did I more passionately wish that God would have intervened somehow to stop this terrible crime. Why didn't He cause the perpetrator to be struck with a bolt of lightning, or at least a pang of conscience? Why didn't He put an invisible shield around my daughter to protect her?

Perhaps on rare occasions God does perform miracles. But for the most part, the existence of free will and of order in the physical universe are inexorable facts. While we might wish for such miraculous deliverance to occur more frequently, the consequence of interrupting these two sets of forces would be utter chaos.

What about the occurrence of natural disasters: earthquakes, tsunamis, volcanoes, great floods and famines? On a smaller but no less poignant scale, what about the occurrence of disease in an innocent victim, such as cancer in a child? The Anglican priest and distinguished physicist John Polkinghorne has referred to this category of event as "physical evil," as opposed to the "moral evil" committed by humankind. How can it be justified?

Science reveals that the universe, our own planet, and life itself are engaged in an evolutionary process. The consequences of that can include the unpredictability of the weather, the slippage of a tectonic plate, or the misspelling of a cancer gene in the normal process of cell division. If at the beginning of time God chose to use these forces to create human beings, then the inevitability of these other painful consequences was also assured. Frequent miraculous interventions would be at least as chaotic in the physical realm as they would be in interfering with human acts of free will.

For many thoughtful seekers, these rational explanations fall short of providing a justification for the pain of human existence. Why is our life more a vale of tears than a garden of delight? Much has been written about this apparent paradox, and the conclusion is not an easy one: if God is loving and wishes the best for us, then perhaps His plan is not the same as our plan. This is a hard concept, especially if we have been too regularly spoon-fed a version of God's benevolence that implies nothing more on His part than a desire for us to be perpetually happy. Again from Lewis: "We want, in fact, not so much a father in Heaven as a grandfather in Heaven—a senile benevolence who, as they say, 'likes to see young people enjoying themselves,' and whose plan for the universe was simply that it might be truly said at the end of each day, 'a good time was had by all.' "[10]

Judging by human experience, if one is to accept God's loving-kindness, He apparently desires more of us than this. Is that not, in fact, your own experience? Have you learned more about yourself when things were going well, or when you were faced with challenges, frustrations, and suffering? "God whispers to us in our pleasures, speaks in our conscience, but shouts in our pains: it is His megaphone to rouse a deaf world."[11] As much as we would like to avoid those experiences, without them would we not be shallow, self-centered creatures who would ultimately lose all sense of nobility or striving for the betterment of others?

Consider this: if the most important decision we are to make on this earth is a decision about belief, and if the most important relationship we are to develop on this earth is a relationship with God, and if our existence as spiritual creatures is not limited to what we can know and observe during our earthly lifetime, then human sufferings take on a wholly new context. We may never fully understand the reasons for these painful experiences, but we can begin to accept the idea that there may be such reasons. In my case I can see, albeit dimly, that my daughter's rape was a challenge for me to try to learn the real meaning of forgiveness in a terribly wrenching circumstance. In complete honesty, I am still working on that. Perhaps this was also an opportunity for me to recognize that I could not truly protect my daughters from all pain and suffering; I had to learn to entrust them to God's loving care, knowing that this provided not an immunization from evil, but a reassurance that their suffering would not be in vain. Indeed, my daughter would say that this experience provided her with the opportunity and motivation to counsel and comfort others who have gone through the same kind of assault.

This notion that God can work through adversity is not an

easy concept, and can find firm anchor only in a worldview that embraces a spiritual perspective. The principle of growth through suffering is, in fact, nearly universal in the world's great faiths. The Four Noble Truths of the Buddha in the Deer Park sermon, for example, begin with "Life is suffering." For the believer, this realization can paradoxically be a source of great comfort.

That woman I cared for as a medical student, for instance, who challenged my atheism with her gentle acceptance of her own terminal illness, saw in this final chapter of her life an experience that brought her closer to God, not further away. On a larger historical stage, Dietrich Bonhoeffer, the German theologian who voluntarily returned to Germany from the United States during World War II to do what he could to keep the real church alive at a time when the organized Christian church in Germany had chosen to support the Nazis, was imprisoned for his role in a plot to assassinate Hitler. During his two years in prison, suffering great indignities and loss of freedom, Bonhoeffer never wavered in his faith or his praise for God. Shortly before he was hanged, only three weeks before the liberation of Germany, he wrote these words: "Time lost is time when we have not lived a full human life, time unenriched by experience, creative endeavor, enjoyment, and suffering."[12]

How Can a Rational Person Believe in Miracles?

Finally, consider an objection to belief that cuts particularly sharply for a scientist. How can miracles be reconciled to a scientific worldview?

In modern parlance, we have cheapened the significance of the word "miracle." We speak of "miracle drugs," "miracle diets," "Miracle on Ice," or even the "miracle Mets." But of course, that's not the original intended meaning of the word. More accurately, a miracle is an event that appears inexplicable by the laws of nature and so is held to be supernatural in origin.

All religions include a belief in certain miracles. The crossing of the Israelites through the Red Sea, led by Moses and accompanied by the drowning of Pharaoh's men, is a powerful story, told in the book of Exodus, of God's providence in preventing the imminent destruction of His people. Similarly, when Joshua asked God to prolong the daylight in order for a particular battle to be successfully carried out, the sun was said to stand still in a way that could only be described as miraculous.

In Islam, the writing of the Qur'an was started in a cave near Mecca, with the instruction of Muhammad provided supernaturally by the angel Jibril. Muhammad's ascension is clearly also a miraculous event, as he is given the opportunity to see all of the features of heaven and hell.

Miracles play a particularly powerful role in Christianity—especially the most significant miracle of all, Christ's rising from the dead.

How can one accept such claims, while claiming to be a rational modern human being? Well, clearly, if one starts out with the presumption that supernatural events are impossible, then no miracles can be allowed. Again, we can turn to C. S. Lewis for particularly clear thinking on this topic, in his book *Miracles*. "Every event which might be claimed to be a miracle is, in the last resort, something presented to our senses, something seen, heard, touched, smelled, or tasted. And our senses are not infallible. If anything extraordinary seems to have happened, we can always say that we have been the victims of an illusion. If we hold a philosophy which excludes the supernatural, this is what we always shall say. What we learn from experience depends on the kind of philosophy we bring to experience. It is therefore useless to appeal to experience before we have settled as well as we can, the philosophical question."[13]

At the risk of frightening those who are uncomfortable with mathematical approaches to philosophical problems, consider the following analysis. The Reverend Thomas Bayes was a Scottish theologian little remembered for his theological musings but much respected for putting forward a particular probability theorem. Bayes's Theorem provides a formula by which one can calculate the probability of observing a particular event, given some initial information (the "prior") and some additional information (the "conditional"). His theorem is particularly useful when facing two or more possible explanations for the occurrence of an event.

Consider the following example. You have been taken captive by a madman. He gives you a chance to be set free—he will allow you to draw a card from a deck, replace it, shuffle, and draw again. If you draw the ace of spades both times, you will be released.

Skeptical of whether this is even worth attempting, you proceed—and to your amazement you draw the ace of spades twice in a row. Your chains are released and you return home.

Being mathematically inclined, you calculate the chances of this good fortune as 1/52 X 1/52 = 1/2704. A very unlikely event, but it happened. A few weeks later, however, you find out that a benevolent employee of the company that manufactured the playing cards, being aware of the madman's wager, had arranged to have one of every hundred decks of cards be made up of fifty-two aces of spades.

So perhaps this was not just a lucky break? Perhaps a knowledgeable and loving being (the employee), unknown to you at the time of your capture, intervened to improve the chances of your release. The likelihood that the deck you drew from was a regular deck of fifty-two different cards was 99/100; the likelihood of a special deck of only aces of spades was 1/100. For those two possible starting points, the "conditional" probabilities of drawing two aces of spades in a row would be 1/2704 and 1, respectively. By Bayes's Theorem it is now possible to calculate the "posterior" probabilities, and conclude that there is a 96 percent likelihood that the deck of cards you drew from was one of the "miraculous" ones.

This same analysis can be applied to apparently miraculous events in daily experience. Suppose you have observed a spontaneous cure of a cancer in an advanced stage, which is known to be fatal in nearly every instance. Is this a miracle? To evaluate that question in the Bayesian sense will require you to postulate what the "prior" is of a miraculous cure of cancer occurring in the first place. Is it one in a thousand? One in a million? Or is it zero?

This is, of course, where reasonable people will disagree, sometimes noisily. For the committed materialist, no allowance can be permitted for the possibility of miracles in the first place (his "prior" will be zero), and therefore even an extremely unusual cure of cancer will be discounted as evidence of the miraculous, and will instead be chalked up to the fact that rare events will occasionally occur within the natural world. The believer in the existence of God, however, may after examining the evidence conclude that no such cure should have occurred by any known natural processes, and having once admitted that the prior probability of a miracle, while quite small, is not quite zero, will carry out his own (very informal) Bayesian calculation to conclude that a miracle is more likely than not.

All of this simply goes to say that a discussion about the miraculous quickly devolves to an argument about whether or not one is willing to consider any possibility whatsoever of the supernatural. I believe that possibility exists, but at the same time, the "prior" should generally be very low. That is, the presumption in any given case should be for a natural explanation. Surprising but mundane events are not automatically miraculous. For the deist, who sees God as having created the universe but then wandering off in some other place to carry out other activities, there is no more reason to consider natural events as miraculous than there is for the committed materialist. For the theist, who believes in a God who is involved in the lives of human beings, various thresholds of assumption of the miraculous are likely to apply, depending on that individual's perception about how likely it is that God would intervene in everyday circumstances.

Whatever the personal view, it is crucial that a healthy skepticism be applied when interpreting potentially miraculous events, lest the integrity and rationality of the religious perspective be brought into question. The only thing that will kill the possibility of miracles more quickly than a committed materialism is the claiming of miracle status for everyday events for which natural explanations are readily at hand. Anyone who claims the blooming of a flower is a miracle is treading upon a growing understanding of plant biology, which is well on the way to elucidating all the steps between seed germination and the blossoming of a beautiful and sweet-smelling rose, all directed by that plant's DNA instruction book.

Similarly, the individual who wins the lottery and announces that this is a miracle, because he prayed about the outcome, strains our credulity. After all, given the wide distribution of at least some vestiges of faith in our modern society, it is likely that a significant fraction of the individuals who bought a lottery ticket that week also prayed in some fleeting way that they might be the winner. If that be so, then the actual winner's claim of miraculous intervention rings hollow.

More difficult to evaluate are the claims of miraculous healing from medical problems. As a physician, I have occasionally seen circumstances where individuals recovered from illnesses that appeared not to be reversible. Yet I am loath to ascribe those events to miraculous intervention, given our incomplete understanding of illness and how it affects the human body. All too often, when claims of miraculous healing have been carefully investigated by objective observers, those claims have fallen short. Despite those misgivings, and an insistence that such claims be backed up by extensive evidence, I would not be stunned to hear that such genuine miraculous healings do occur on extremely rare occasions. My "prior" is low, but it is not zero.

Miracles thus do not pose an irreconcilable conflict for the believer who trusts in science as a means to investigate the natural world, and who sees that the natural world is ruled by laws. If, like me, you admit that there might exist something or someone outside of nature, then there is no logical reason why that force could not on rare occasions stage an invasion. On the other hand, in order for the world to avoid descending into chaos, miracles must be very uncommon. As Lewis has written, "God does not shake miracles into nature at random as if from a pepper-caster. They come on great occasions: they are found at the great ganglions of history—not of political or social history, but of that spiritual history which cannot be fully known by men. If your own life does not happen to be near one of those great ganglions, how should you expect to see one?"[14]

Here we see not only an argument about the rarity of miracles, but an argument that they should have some purpose, rather than representing the supernatural acts of a capricious magician, simply designed to amaze. If God is the ultimate embodiment of omnipotence and goodness, He would not play such a trickster role. John Polkinghorne argues this point cogently: "Miracles are not to be interpreted as divine acts against the laws of nature (for those laws are themselves expressions of God's will) but as more profound revelations of the character of the divine relationship to creation. To be credible, miracles must convey a deeper understanding than could have been obtained without them."[15]

Despite these arguments, materialistic skeptics who wish to give no ground to the concept of the supernatural, those who refute the evidence from the Moral Law and the universal sense of longing for God, will no doubt argue that there is no need to consider miracles at all. In their view, the laws of nature can explain everything, even the exceedingly improbable.

But can this view be completely sustained? There is at least one singular, exceedingly improbable, and profound event in history that scientists of nearly all disciplines agree is not understood and will never be understood, and for which the laws of nature fall completely short of providing an explanation. Would that be a miracle? Read on.

PART TWO

The Great Questions of Human Existence

CHAPTER THREE

The Origins of the Universe

MORE THAN TWO HUNDRED YEARS AGO, one of the most influential philosophers of all time, Immanuel Kant, wrote: "Two things fill me with constantly increasing admiration and awe, the longer and more earnestly I reflect on them: the starry heavens without and the Moral Law within." An effort to understand the origins and workings of the cosmos has characterized nearly all religions throughout history, whether in the overt worship of a sun god, the ascription of spiritual significance to phenomena such as eclipses, or simply a sense of awe at the wonders of the heavens.

Was Kant's remark merely the sentimental musing of a philosopher not benefited by discoveries of modern science, or is there a harmony achievable between science and faith in the profoundly important question of the origins of the universe?

One of the challenges in achieving that harmony is that science is not static. Scientists are constantly reaching into new arenas, investigating the natural world in new ways, digging deeper into territory where understanding is incomplete. Faced with a set of data that includes a puzzling and unexplained phenomenon, scientists construct hypotheses of the mechanism that might be involved, and then conduct experiments to test those hypotheses. Many experiments on the cutting edge of science fail, and most hypotheses turn out to be wrong. Science is progressive and self-correcting: no significantly erroneous conclusions or false hypotheses can be sustained for long, as newer observations will ultimately knock down incorrect constructs. But over a long period of time, a consistent set of observations sometimes emerges that leads to a new framework of understanding. That framework is then given a much more substantive description, and is called a "theory"—the theory of gravitation, the theory of relativity, or the germ theory, for instance.

One of the most cherished hopes of a scientist is to make an observation that shakes up a field of research. Scientists have a streak of closeted anarchism, hoping that someday they will turn up some unexpected fact that will force a disruption of the framework of the day. That's what Nobel Prizes are given for. In that regard, any assumption that a conspiracy could exist among scientists to keep a widely current theory alive when it actually contains serious flaws is completely antithetical to the restless mind-set of the profession.

The study of astrophysics nicely exemplifies these principles. Profound upheavals have occurred over the last five hundred years, during which the understanding of the nature of matter and the structure of the universe has undergone major revisions. No doubt more revisions still lie ahead of us.

These disruptions can sometimes be wrenching for attempts to achieve a comfortable synthesis between science and faith, especially if the church has attached itself to a prior view of things and incorporated that into its core belief system. Today's harmony can be tomorrow's discord. In the sixteenth and seventeenth centuries, Copernicus, Kepler, and Galileo (all strong believers in God) built an increasingly compelling case that the movement of the planets could be properly understood only if the earth revolved around the sun, rather than the other way around. The details of their conclusions were not all quite correct (Galileo made a famous blooper in his explanation of the tides), and many in the scientific community were initially unconvinced, but ultimately the data and the consistency of the theory's predictions convinced even the most skeptical scientists. The Catholic Church remained strongly opposed, however, claiming that this view was incompatible with holy scripture. In retrospect it is clear that the scriptural basis for those claims was remarkably thin; nonetheless, this confrontation raged for decades and ultimately did considerable harm, both to science and to the church.

The past century has seen an unprecedented number of revisions in our view of the universe. Matter and energy, previously assumed to be utterly different entities, were shown by Einstein to be interchangeable by the famous equation $E = mc^2$ (*E* is energy, *m* is mass, and *c* is the speed of light). The dualism of wave and particle—that is, the fact that matter has simultaneous characteristics of both waves and particles—a phenomenon demonstrated experimentally for light and small particles such as electrons, was unanticipated and astounding to many classically trained scientists. The Heisenberg uncertainty principle of quantum mechanics, the realization that it is possible to measure either the position or the momentum of a particle, but not both at once, created particularly disruptive consequences for both science and theology. Perhaps most profoundly, our concept of the origin of the universe has undergone a fundamental change over the course of the past seventy-five years, on the basis of both theory and experiment.

Most of these massive revisions in our understanding of the material universe have come about within relatively narrow circles of academic investigation, and have remained largely out of view for the general public. Occasional noble efforts, such as Stephen Hawking's *A Brief History of Time*, have attempted to explain the complexities of modern physics and cosmology to a more general audience, but it seems likely that the 5 million printed copies of Hawking's book remain largely unread by an audience that overwhelmingly found the concepts within its pages just too bizarre to comprehend.

Indeed, discoveries about physics in the last few decades have led to insights about the nature of matter that are profoundly counterintuitive. The physicist Ernest Rutherford commented one hundred years ago that "a theory that you can't explain to a bartender is probably no damn good." By this standard, many of the current theories about the fundamental particles that make up all matter hold up rather poorly.

Among the many strange concepts now well documented experimentally are such things as the fact that neutrons and protons (which we used to think were *the* fundamental particles of the atomic nucleus) are actually made up of six flavors of quarks (named "up," "down," "strange," "charmed," "bottom," and "top"). The six flavors become even stranger when they are described as each having three colors (red, green, and blue). The quirky names given to these particles at least prove that scientists have a sense of humor. A dizzying array of other particles, from photons to gravitons to gluons and muons, create a world so foreign to everyday human experience that they cause many nonscientists to shake their heads in disbelief. Yet all of these particles make possible our very existence. For those who argue that materialism should be favored over theism, because materialism is simpler and more intuitive, these new concepts present a major challenge. A variation on Ernest Rutherford's dictum is famously known as Occam's Razor, a misspelled attribution to the fourteenth-century English logician and monk William of Ockham. This principle suggests that the simplest explanation for any given problem is usually best. Today, Occam's Razor appears to have been relegated to the Dumpster by the bizarre models of quantum physics.

But in one very important sense, Rutherford and Occam are still honored: as puzzling as the verbal descriptions of these newly discovered phenomena are, their mathematical representation invariably turns out to be elegant, unexpectedly simple, and even beautiful. When I was a graduate student in physical chemistry at Yale, I had the remarkable experience of taking a course in relativistic quantum mechanics from Nobel laureate Willis Lamb. His class style was to work through the theories of relativity and quantum mechanics from first principles. He did this entirely from memory but occasionally skipped steps and charged us, his wide-eyed student admirers, to fill in the gaps before coming to the next class.

Though I ultimately moved on from physical science to biology, this experience of deriving simple and beautiful universal equations that describe the reality of the natural world left a profound impression on me, particularly because the ultimate outcome had such aesthetic appeal. This raises the first of several philosophical questions about the nature of the physical universe. Why should matter behave in such a way? In Eugene Wigner's phrase, what could be the explanation for the "unreasonable effectiveness of mathematics"?[1]

Is this no more than a happy accident, or does it reflect some profound insight into the nature of reality? If one is willing to accept the possibility of the supernatural, is it also an insight into the mind of God? Were Einstein, Heisenberg, and others encountering the divine?

In the final sentences of *A Brief History of Time,* referring to a hoped-for time when an eloquent and unified theory of everything is developed, Stephen Hawking (not generally given to metaphysical musings) says, "Then we shall all, philosophers, scientists, and just ordinary people, be able to take part in the discussion of the question of why it is that we and the universe exist. If we find the answer to that, it would be the ultimate triumph of human reason—for then we would know the mind of God."[2] Are these mathematical descriptions of reality signposts to some greater intelligence? Is mathematics, along with DNA, another language of God?

Certainly, mathematics has led scientists right to the doorstep of some of the most profound questions of all. First among them: how did it all begin?

THE BIG BANG

At the beginning of the twentieth century, most scientists assumed a universe with no beginning and no end. This created certain physical paradoxes, such as how the universe managed to remain stable without collapsing upon itself because of the force of gravity, but other alternatives did not seem very attractive. When Einstein developed the theory of general relativity in 1916, he introduced a "fudge factor" to block gravitational implosion and retain the idea of a steady-state universe. He later reportedly called this "the greatest mistake of my life."

Other theoretical formulations proposed the alternative of a universe that had begun at a particular moment, and then expanded to its present state; but it remained for experimental measurements to confirm this before most physicists were willing to consider that hypothesis seriously. Those data were initially provided by Edwin Hubble in 1929, in a famous set of experiments in which he looked at the rate at which neighboring galaxies are receding from our own.

Using the Doppler effect—the same principle that allows the state police to determine the speed of your car as you pass by their radar equipment, or that causes the whistle of an oncoming train to have a higher pitch than after it has passed you—Hubble found that everywhere he looked, the light in the galaxies suggested that they were receding from ours. The farther away they were, the faster the galaxies were receding.

If everything in the universe is flying apart, reversing the arrow of time would predict that at some point all of these galaxies were together in one incredibly massive entity. Hubble's observations started a deluge of experimental measurements that over the last seventy years have led to the conclusion by the vast majority of physicists and cosmologists that the universe began at a single moment, commonly now referred to as the Big Bang. Calculations suggest it happened approximately 14 billion years ago.

A particularly important documentation of the correctness of this theory was provided rather accidentally by Arno Penzias and Robert Wilson in 1965, when they detected what appeared to be an annoying background of microwave signals regardless of where they pointed their new detector. After ruling out all other possible causes (including certain pigeons, who were initially suspected as the culprits), Penzias and Wilson ultimately realized that this background noise was coming from the universe itself, and that it represented precisely the kind of afterglow that one would expect to find as a consequence of the Big Bang, arising from the annihilation of matter and antimatter in the early moments of the exploding universe.

Additional compelling evidence for the correctness of the Big Bang theory has been provided by the ratio of certain elements throughout the universe, particularly hydrogen, deuterium, and helium. The abundance of deuterium is remarkably constant, from nearby stars to the farthest-flung galaxies near our event horizon. That finding is consistent with all of the universe's deuterium having been formed at unbelievably high temperatures in a single event during the Big Bang. If there were multiple such events in different locations and times, we would not expect such uniformity.

Based on these and other observations, physicists are in agreement that the universe began as an infinitely dense, dimensionless point of pure energy. The laws of physics break down in this circumstance, referred to as a "singularity." At least so far, scientists have been unable to interpret the very earliest events in the explosion, occupying the first 10^{-43} seconds (one tenth of a millionth of a millionth of a millionth of a millionth of a millionth of a millionth of a second!). After that, it is possible to make predictions about the events that would need to have occurred to result in today's observable universe, including the annihilation of matter and antimatter, the formation of stable atomic nuclei, and ultimately the formation of atoms, primarily hydrogen, deuterium, and helium.

A currently unanswered question is whether the Big Bang has resulted in a universe that will go on expanding forever, or whether at some point gravitation will take over and the galaxies will begin to fall back together, ultimately resulting in a "Big Crunch." Recent discoveries of little-understood quantities known as dark matter and dark energy, which seem to occupy a very substantial amount of the material in the universe, leave the answer to this question hanging, but the best evidence at the moment predicts a slow fade, rather than a dramatic collapse.

WHAT CAME BEFORE THE BIG BANG?

The existence of the Big Bang begs the question of what came before that, and who or what was responsible. It certainly demonstrates the limits of science as no other phenomenon has done. The consequences of Big Bang theory for theology are profound. For faith traditions that describe the universe as having been created by God from nothingness (ex nihilo), this is an electrifying outcome. Does such an astonishing event as the Big Bang fit the definition of a miracle?

The sense of awe created by these realizations has caused more than a few agnostic scientists to sound downright theological. In *God and the Astronomers,* the astrophysicist Robert Jastrow wrote this final paragraph: "At this moment it seems as though science will never be able to raise the curtain on the mystery of creation. For the scientist who has lived by his faith in the power of reason, the story ends like a bad dream. He has scaled the mountains of ignorance; he is about to conquer the highest peak; as he pulls himself over the final rock, he is greeted by a band of theologians who have been sitting there for centuries."[3]

For those looking to bring the theologians and the scientists closer together, there is much in these recent discoveries of the origin of the universe to inspire mutual appreciation. Elsewhere in his provocative book, Jastrow writes: "Now we see how the astronomical evidence leads to a biblical view of the origin of the world. The details differ, but the essential elements and the astronomical and biblical accounts of Genesis are the same; the chain of events leading to man commenced suddenly and sharply at a definite moment in time, in a flash of light and energy."[4]

I have to agree. The Big Bang cries out for a divine explanation. It forces the conclusion that nature had a defined beginning. I cannot see how nature could have created itself. Only a supernatural force that is outside of space and time could have done that.

But what of the rest of creation? What are we to make of the long, drawn-out process by which our own planet, Earth, came into existence, some 10 billion years after the Big Bang?

FORMATION OF OUR SOLAR SYSTEM AND PLANET EARTH

For the first million years after the Big Bang, the universe expanded, the temperature dropped, and nuclei and atoms began to form. Matter began to coalesce into galaxies under the force of gravity. It acquired rotational motion as it did so, ultimately resulting in the spiral shape of galaxies such as our own. Within those galaxies local collections of hydrogen and helium were drawn together, and their density and temperature rose. Ultimately nuclear fusion commenced.

This process, whereby four hydrogen nuclei fuse together to form both energy and a helium nucleus, provides the major source of fuel for stars. Larger stars burn faster. As they begin to burn out, they generate within their core even heavier elements such as carbon and oxygen. Early in the universe (within the first few hundred million years) such elements appeared only in the core of these collapsing stars, but some of these stars then went through massive explosions known as supernovae, flinging heavier elements back into the gas in the galaxy.

Scientists believe our own sun did not form in the early days of the universe; our sun is instead a second- or third-generation star, formed about 5 billion years ago by a local re-coalescence. As that was occurring, a small proportion of heavier elements in the vicinity escaped incorporation into the new star, and instead collected into the planets that now rotate around our sun. This includes our own planet, which was far from hospitable in its early days. Initially very hot, and bombarded with continual massive collisions, Earth gradually cooled, developed an atmosphere, and became potentially hospitable to living things by about 4 billion years ago. A mere 150 million years later, the earth was teeming with life.

All of these steps in the formation of our solar system are now well described and unlikely to be revised on the basis of additional future information. Nearly all of the atoms in your body were once cooked in the nuclear furnace of an ancient supernova—you are truly made of stardust.

Are there theological implications to any of these discoveries? How rare are we? How unlikely?

An argument can be made that the origin of complex life forms in this universe could not have happened in less than about 5–10 billion years after the Big Bang, since the first generation of stars would not have contained the heavier elements like carbon and oxygen that we believe are necessary for life, at least as we know it. Only a second- or third-generation star, and its accompanying planetary system, would carry that potential. Even then, a great deal of time would be necessary for life to reach sentience and intelligence. While other life forms not dependent on heavy elements might potentially exist elsewhere in the universe, the nature of such organisms is extremely difficult to contemplate from our current knowledge of chemistry and physics.

This does, of course, raise the question about whether life exists elsewhere in the universe of a sort that we would recognize. While no one on earth has any current data to support or refute this, a famous equation proposed by radio astronomer Frank Drake in 1961 allowed a consideration of what the probabilities might be. The Drake equation is most useful as a way of documenting the state of our ignorance. Drake noted, simply and logically, that the number of communicating civilizations in our own galaxy must be the product of seven factors:

* the number of stars in the Milky Way galaxy (about 100 billion), *times*
* the fraction of stars that have planets around them, *times*
* the number of planets per star that are capable of sustaining life, *times*
* the fraction of those planets where life actually evolves, *times*

* the fraction of these where the life that evolves is intelligent, *times*
* the fraction of these that actually developed the ability to communicate, *times*
* the fraction of these planets' life during which the ability to communicate overlaps with ours

We have been able to communicate beyond Earth for less than a hundred years. The earth is approximately 4.5 billion years old, so Drake's last factor reflects only a tiny fraction of Earth's years of existence: 0.000000022. (One might argue, depending on one's perspective about the distinct likelihood of our destroying ourselves in the future, whether that fraction will ever get much larger than this.)

Drake's formula is interesting but essentially useless, because of our inability to state with any degree of certainty the value of almost all of the terms except for the number of stars in the Milky Way galaxy. Certainly other stars have been discovered with planets around them, but the rest of the terms remain hidden in mystery. Nonetheless, the Search for Extraterrestrial Intelligence (SETI) Institute, founded by Frank Drake himself, has now engaged amateur and professional physicists, astronomers, and others in an organized effort to seek signals that might be coming from other civilizations in our galaxy.

Much has been written about the potential theological significance of the discovery of life on other planets, should that happen to come to pass. Would such an event automatically render humankind on planet Earth less "special"? Would the existence of life on other planets make a creator God involved in the process less likely? In my view, such conclusions do not really seem warranted. If God exists, and seeks to have fellowship with sentient beings like ourselves, and can handle the challenge of interacting with 6 billion of us currently on this planet and countless others who have gone before, it is not clear why it would be beyond His abilities to interact with similar creatures on a few other planets or, for that matter, a few million other planets. It would, of course, be of great interest to discover whether such creatures in other parts of the universe also possess the Moral Law, given its importance in our own perception of the nature of God. Realistically, however, it is unlikely that any of us will have the opportunity to learn the answers to those questions during our lifetime.

THE ANTHROPIC PRINCIPLE

Now that the origin of the universe and our own solar system has become increasingly well understood, a number of fascinating apparent coincidences about the natural world have been discovered that have puzzled scientists, philosophers, and theologians alike. Consider the following three observations:

> 1. In the early moments of the universe following the Big Bang, matter and antimatter were created in almost equivalent amounts. At one millisecond of time, the universe cooled enough

71

for quarks and antiquarks to "condense out." Any quark encountering an antiquark, which would happen quickly at this high density, resulted in the complete annihilation of both and the release of a photon of energy. But the symmetry between matter and antimatter was not quite precise; for about every billion pair of quarks and antiquarks, there was an extra quark. It is that tiny fraction of the initial potentiality of the entire universe that makes up the mass of the universe as we now know it.

Why did this asymmetry exist? It would seem more "natural" for there to be no asymmetry. But if there had been complete symmetry between matter and antimatter, the universe would quickly have devolved into pure radiation, and people, planets, stars, and galaxies would never have come into existence.

2. The way in which the universe expanded after the Big Bang depended critically on how much total mass and energy the universe had, and also on the strength of the gravitational constant. The incredible degree of fine-tuning of these physical constants has been a subject of wonder for many experts. Hawking writes: "Why did the universe start out with so nearly the critical rate of expansion that separates models that recollapse from those that go on expanding forever, that even now, 10 thousand million years later, it is still expanding at nearly the critical rate? If the rate of ex-

pansion one second after the Big Bang had been smaller by even one part in 100 thousand million million, the universe would have recollapsed before it ever reached its present size."[5]

On the other hand, if the rate of expansion had been greater by even one part in a million, stars and planets could not have been able to form. Recent theories involving an incredibly rapid expansion (inflation) of the universe at very early times appear to offer a partial explanation for why the present expansion is so close to the critical value. However, many cosmologists would say that this simply pushes the question back to why the universe had just the right properties to undergo such an inflationary expansion. The existence of a universe as we know it rests upon a knife edge of improbability.

3. The same remarkable circumstance applies to the formation of heavier elements. If the strong nuclear force that holds together protons and neutrons had been even slightly weaker, then only hydrogen could have formed in the universe. If, on the other hand, the strong nuclear force had been slightly stronger, all the hydrogen would have been converted to helium, instead of the 25 percent that occurred early in the Big Bang, and thus the fusion furnaces of stars and their ability to generate heavier elements would never have been born.

Adding to this remarkable observation, the nuclear force appears to be tuned just sufficiently for carbon to form, which is critical for life forms on Earth. Had that force been just slightly more attractive, all the carbon would have been converted to oxygen.

Altogether, there are fifteen physical constants whose values current theory is unable to predict. They are givens: they simply have the value that they have. This list includes the speed of light, the strength of the weak and strong nuclear forces, various parameters associated with electromagnetism, and the force of gravity. The chance that all of these constants would take on the values necessary to result in a stable universe capable of sustaining complex life forms is almost infinitesimal. And yet those are exactly the parameters that we observe. In sum, our universe is wildly improbable.

You may rightly object at this point that this argument is a bit circular: the universe had to have parameters associated with this kind of stability or we would not be here to comment upon it. This general conclusion is referred to as the Anthropic Principle: the idea that our universe is uniquely tuned to give rise to humans. It has been a source of much wonder and speculation since it was fully appreciated a few decades ago.[6]

Essentially, there are three possible responses to the Anthropic Principle:

> 1. There may be an essentially infinite number of universes, either occurring simultaneously with our own or in some sequence, with different val-

ues of the physical constants, and maybe even different physical laws. We are, however, unable to observe the other universes. We can exist only in a universe where all the physical properties work together to permit life and consciousness. Ours is not miraculous, it is simply an unusual product of trial and error. This is called the "multiverse" hypothesis.

2. There is only one universe, and this is it. It just happened to have all the right characteristics to give rise to intelligent life. If it hadn't, we wouldn't be here discussing this. We are just very, very, very lucky.

3. There is only one universe, and this is it. The precise tuning of all of the physical constants and physical laws to make intelligent life possible is not an accident, but reflects the action of the one who created the universe in the first place.

Regardless of one's preference for option 1, 2, or 3, there is no question that this is potentially a theological issue. Hawking, quoted by Ian Barbour,[7] writes, "The odds against a universe like ours emerging out of something like the Big Bang are enormous. I think there are clearly religious implications."

Going even further, in *A Brief History of Time,* Hawking states: "It would be very difficult to explain why the universe should have begun in just this way, except as the act of a God who intended to create beings like us."[8]

Another distinguished physicist, Freeman Dyson, after reviewing this series of "numerical accidents," concludes, "The more I examine the universe and the details of its architecture, the more evidence I find that the universe in some sense must have known we were coming."[9] And Arno Penzias, the Nobel Prize–winning scientist who codiscovered the cosmic microwave background radiation that provided strong support for the Big Bang in the first place, states, "The best data we have are exactly what I would have predicted, had I nothing to go on but the five Books of Moses, the Psalms, the Bible as a whole."[10] Perhaps Penzias was thinking of the words of David in Psalm 8: "When I consider your heavens, the work of your fingers, the moon and the stars, which you have set in place, what is man that you are mindful of him?"

So where should we come down on the three options listed above? Let's approach it logically. To begin with, we have the observation of the universe as we know it, including ourselves. We then wish to calculate which of these three possible options is most likely. The problem is, we don't have a good way of deciding the landscape of probabilities, except perhaps for option 2. For option 1, as the number of parallel universes approaches infinity, then the likelihood of at least one of them having the physical properties for life could be substantial. For option 2, however, the probability will be vanishingly small. The likelihood of option 3 depends on the existence of a supernatural Creator who cares about a nonsterile universe.

On the basis of probability, option 2 is the least plausible. That then leaves us with option 1 and option 3. The first is logically defensible, but this near-infinite number of unobservable universes strains credulity. It certainly fails Occam's Razor.

Those categorically unwilling to accept an intelligent Creator will argue, however, that option 3 is not simpler at all, since it requires the intercession of a supernatural being. It could be argued, however, that the Big Bang *itself* seems to point strongly toward a Creator, since otherwise the question of what came before is left hanging in the air.

If one is willing to accept the argument that the Big Bang requires a Creator, then it is not a long leap to suggest that the Creator might have established the parameters (physical constants, physical laws, and so on) in order to accomplish a particular goal. If that goal happened to include a universe that was more than a featureless void, then we have arrived at option 3.

In trying to judge between options 1 and 3, a particular parable by philosopher John Leslie comes to mind.[11] In this parable, an individual faces a firing squad, and fifty expert marksmen aim their rifles to carry out the deed. The order is given, the shots ring out, and yet somehow all of the bullets miss and the condemned individual walks away unscathed.

How could such a remarkable event be explained? Leslie suggests there are two possible alternatives, which correspond to our options 1 and 3. In the first place, there may have been thousands of executions being carried out in that same day, and even the best marksmen will occasionally miss. So the odds just happen to be in favor of this one individual, and all fifty of the marksmen fail to hit the target. The other option is that something more directed is going on, and the apparent poor aim of the fifty experts was actually intentional. Which seems more plausible?

One must leave open the door to the possibility that future investigation in theoretical physics will demonstrate that some of the fifteen physical constants that so far are simply determined by experimental observation may be limited in their potential numerical value by something more profound, but such a revelation is not currently on the horizon. Furthermore, as with other arguments in this chapter and those that precede and follow it, no scientific observation can reach the level of absolute proof of the existence of God. But for those willing to consider a theistic perspective, the Anthropic Principle certainly provides an interesting argument in favor of a Creator.

QUANTUM MECHANICS AND THE UNCERTAINTY PRINCIPLE

Isaac Newton was a believer who wrote more about biblical interpretation than he did about mathematics and physics, but not all those who followed him shared that same faith. At the beginning of the nineteenth century, the marquis de Laplace, a distinguished French mathematician and physicist, put forward the point of view that nature is governed by a set of precise physical laws (some discovered, some yet to be discovered), and nature is therefore unable to avoid adhering to those laws. In Laplace's view, that requirement would extend to the tiniest particles, the most far-flung parts of the universe, and also to human beings and their thought processes.

Laplace postulated that once the initial configuration of the universe was established, all other future events, including those involving human experiences of the past, present, and future, were irreversibly specified. This represented an extreme form of scientific determinism, obviously leaving no place for God (except at the beginning) or the concept of free will. It created quite a stir in the scientific and theological communities. (As Laplace famously said to Napoleon, when asked about God, "I have no need of that hypothesis.")

A century later Laplace's concept of precise scientific determinism was overturned, not by theological arguments but by scientific insights. The revolution known as quantum mechanics began, simply enough, as an effort to explain an unsolved problem in physics concerning the spectrum of light. Based on a number of observations, Max Planck and Albert Einstein demonstrated that light did not come in all possible energies, but that it was "quantized" in particles of precise energy, known as photons. At bottom, therefore, light is not infinitely indivisible, but comprises a flow of photons, just as the resolution of a digital camera cannot be any finer than a single pixel.

At the same time, Niels Bohr examined the structure of the atom and wondered how it is that electrons manage to remain in orbit around the nucleus. The negative charge of each electron should attract it to the positive charge of each proton in the nucleus, ultimately resulting in an unavoidable implosion of all matter. Bohr postulated a similar quantum argument, developing a theory that postulated that electrons could exist only in a certain number of finite states.

The foundations of classical mechanics began to crack, but the most profound philosophical consequences of these revelations appeared subsequently from physicist Werner Heisenberg, when he showed convincingly that in this bizarre quantum

world of very small distances and tiny particles, it was impossible to measure the position *and* momentum of a particle accurately at the same time. This uncertainty principle, which bears
Heisenberg's name, overturned Laplacean determinism in one
stroke, since it indicated that any initial configuration of the
universe could never actually be determined as precisely as
would be required for Laplace's predictive model.

The consequences of quantum mechanics for an understanding of the meaning of the universe have been the subject
of much speculation over the last eighty years. Einstein himself,
though he played an important role in the early development of
quantum mechanics, initially rejected the concept of uncertainty, famously remarking, "God does not play dice."

The theist might reply that the game would not appear to
be dice to God, even if it does to us. As Hawking points out,
"We could still imagine that there is a set of laws that deter-
mines events completely for some supernatural being, who
could observe the present state of the universe without disturbing it."[12]

COSMOLOGY AND THE GOD HYPOTHESIS

This brief review of the nature of the universe leads to a recon-
sideration of the plausibility of the God hypothesis in a more
general way. I am reminded of Psalm 19, where David writes,
"The heavens declare the glory of God; the skies proclaim the
work of His hands." Clearly, the scientific worldview is *not* entirely sufficient to answer all of the interesting questions about

the origin of the universe, and there is nothing inherently in conflict between the idea of a creator God and what science has revealed. In fact, the God hypothesis solves some deeply troubling questions about what came before the Big Bang, and why the universe seems to be so exquisitely tuned for us to be here.

For the theist, who is led from the Moral Law argument (Chapter 1) to seek a God who not only set the universe in motion, but takes an interest in human beings, such a synthesis can be readily achieved. The argument would go something like this:

> If God exists, then He is supernatural.
> If He is supernatural, then He is not limited by natural laws.
> If He is not limited by natural laws, there is no reason He should be limited by time.
> If He is not limited by time, then He is in the past, the present, and the future.

The consequence of those conclusions would include:

> He could exist before the Big Bang and He could exist after the universe fades away, if it ever does.
> He could know the precise outcome of the formation of the universe even before it started.
> He could have foreknowledge of a planet near the outer rim of an average spiral galaxy that

would have just the right characteristics to
allow life.

He could have foreknowledge that that planet
would lead to the development of sentient
creatures, through the mechanism of evolution by natural selection.

He could even know in advance the thoughts and
actions of those creatures, even though they
themselves have free will.

I will have much more to say about the latter steps in this
synthesis, but the outlines of a satisfying harmony between science and belief can now be seen.

This proposed synthesis is not intended to gloss over all
challenges and areas of discord. Believers in particular world
religions certainly encounter specific difficulties with some of
the details of the origin of the universe predicted by science.

Deists like Einstein, who view God as having started the
whole process but then paying no attention to subsequent developments, are generally comfortable with recent conclusions
of physics and cosmology, with the possible exception of the
uncertainty principle. But the comfort level of the major theistic religions is somewhat variable. The idea of a finite beginning of the universe is not entirely resonant with Buddhism,
where an oscillating universe would be more compatible. But
the theistic branches of Hinduism encounter no major conflict
with the Big Bang. Neither do most (but not all) interpreters of
Islam.

For the Judaeo-Christian tradition, the opening words of

Genesis ("In the beginning, God created the heavens and the earth") are entirely compatible with the Big Bang. In one notable example, Pope Pius XII of the Roman Catholic Church was a strong supporter of Big Bang theory even before its scientific underpinnings were well established.

Not all Christian interpretations have been so supportive of this scientific view of the universe, however. Those who interpret Genesis in absolutely literal terms conclude that the earth is only six thousand years old, and therefore reject most of the conclusions just cited. Their position is in some ways understandable as an appeal to truth: believers in a religion that is undergirded by sacred texts rightly object to loose interpretations of their meaning. Texts that seem to describe historical events should be interpreted as allegory only if strong evidence requires it.

But is Genesis in this category? Unquestionably the language is poetic. Does it exhibit poetic license? (There will be much more to say about this in a later chapter.) This is not just a modern-day question; throughout history debates have raged between literalists and nonliteralists. Saint Augustine, probably one of the greatest of all religious intellects, was particularly aware of the risks of turning biblical texts into precise scientific treatises, and wrote, with specific reference to Genesis: "In matters that are so obscure and far beyond our vision, we find in Holy Scripture passages which can be interpreted in very different ways without prejudice to the faith we have received. In such cases, we should not rush in headlong and so firmly take our stand on one side that, if further progress in the search for truth justly undermines this position, we too fall with it."[13]

The next chapters look more closely at those aspects of science devoted to the study of life. The potential conflicts between science and faith, at least as perceived by many modern commentators, will continue to appear. But I will argue that if we wisely apply Saint Augustine's advice, crafted well over a thousand years before there was any reason to be apologetic about Darwin, we will be able to find a consistent and profoundly satisfying harmony between these worldviews.

CHAPTER FOUR

Life on Earth

Of Microbes and Man

THE ADVANCES OF SCIENCE in the modern age have come at the cost of certain traditional reasons for belief in God. When we had no idea how the universe came into existence, it was easier to ascribe it all to an act of God, or many separate acts of God. Similarly, until Kepler, Copernicus, and Galileo upset the applecart in the sixteenth century, the placement of Earth at the center of the majestic starry heavens seemed to represent a powerful argument for the existence of God. If He put us on center stage, He must have built it all for us. When heliocentric science forced a revision of this perception, many believers were shaken up.

But a third pillar of belief continued to carry considerable weight: the complexity of earthly life, implying to any reasonable observer the handiwork of an intelligent designer. As we shall see, science has now turned this upside down. But here, as with the other two arguments, I would like to suggest that science should not be denied by the believer, it should be embraced. The elegance behind life's complexity is indeed reason for awe, and for belief in God—but not in the simple, straightforward way that many found so compelling before Darwin came along.

THE "ARGUMENT FROM DESIGN" dates back at least to Cicero. It was put forward with particular effectiveness by William Paley in 1802 in a highly influential book, *Natural Theology, or Evidences of the Existence and Attributes of the Deity Collected from the Appearance of Nature.* Paley, a moral philosopher and Anglican priest, posed the famous watchmaker analogy:

> In crossing a heath, suppose I pitched my foot against a stone, and were asked how the stone came to be there; I might possibly answer that, for anything I knew to the contrary, it had lain there forever. Nor would it perhaps be very easy to show the absurdity of this answer. But suppose I had found a watch upon the ground, and it should be inquired how the watch happened to be in that place; I should hardly think of the answer, which I had before given, that for anything I knew, the watch might have always been there . . . the watch must have had a maker: that there must have ex-

isted, at some time, and at some place or other, an
artificer or artificers, who formed it for the purpose
which we find it actually to answer; who compre-
hended its construction, and designed its use. . . .
Every indication of contrivance, every manifestation of design, which existed in the watch, exists in
the works of nature; with the difference, on the side
of nature, of being greater or more, and that in a
degree which exceeds all computation.[1]

The evidence of design in nature has been compelling to
humanity throughout much of our existence. Darwin himself,
before his voyage on the HMS *Beagle,* was an admirer of Paley's
writings, and professed to be convinced by this view. However,
even simply as a matter of logic, there is a flaw in Paley's argument. The point he is making can be summarized as follows:

1. A watch is complex.
2. A watch has an intelligent designer.
3. Life is complex.
4. Therefore, life also has an intelligent designer.

But the fact that two objects share one characteristic
(complexity) does not imply they will share all characteristics.
Consider, for example, the following parallel argument:

1. Electric current in my house consists of a flow of
   electrons.
2. Electric current comes from the power company.

3. Lightning consists of a flow of electrons.
4. Therefore, lightning comes from the power company.

As appealing as it seems, Paley's argument cannot be the whole story. To examine the complexity of life and our own origins on this planet, we must dig deep into the fascinating revelations about the nature of living things wrought by the current revolution in paleontology, molecular biology, and genomics. A believer need not fear that this investigation will dethrone the divine; if God is truly Almighty, He will hardly be threatened by our puny efforts to understand the workings of His natural world. And as seekers, we may well discover from science many interesting answers to the question "How does life work?" What we cannot discover, through science alone, are the answers to the questions "Why is there life anyway?" and "Why am I here?"

ORIGINS OF LIFE ON PLANET EARTH

Science begins to answer the question of life's complexity with a timeline. We now know that the universe is approximately 14 billion years old. A century ago, we didn't even know how long our own planet had been around. But the subsequent discovery of radioactivity and the natural decay of certain chemical isotopes provided an elegant and rather precise means of determining the age of various rocks on Earth. The scientific basis of this method is described in detail in Brent Dalrymple's book *The Age of the Earth,* and depends upon the known and very long half-lives by which three radioactive chemical elements steadily decay and transform into different, stable elements: uranium slowly becomes lead, potassium slowly becomes argon, and the more exotic strontium becomes the rare element called rubidium. By measuring the amounts of any of these pairs of elements, we can estimate the age of any particular rock. All of these independent methods give results that are strikingly concordant, pointing to an age of Earth of 4.55 billion years, with an estimated error of only about 1 percent. The oldest rocks that have been dated on the current earth surface are approximately 4 billion years old, but nearly seventy meteorites and a number of moon rocks have been dated at 4.5 billion years old.

All evidence currently available suggests that the earth was a very inhospitable place for its first 500 million years. The planet was under constant and devastating attack from giant asteroids and meteorites, one of which actually tore the moon loose from Earth. Not surprisingly, therefore, rocks dating back 4 billion years or more show absolutely no evidence of any life forms. Just 150 million years later, however, multiple different types of microbial life are found. Presumably, these single-celled organisms were capable of information storage, probably using DNA, and were self-replicating and capable of evolving into multiple different types.

Recently, Carl Woese has put forward the plausible hypothesis that at this particular time on earth, exchange of DNA between organisms was readily accomplished.[2] Essentially, the biosphere consisted of a large number of minuscule independent cells, but they interacted extensively with one another. If a particular organism developed a protein or series of proteins that provided a certain advantage, those new features could quickly be acquired by its neighbors. Perhaps in that sense, early evolution was more a communal than an individual activity. This kind of "horizontal gene transfer" is well documented in the most ancient forms of bacteria that now exist on the planet (archaebacteria), and may have provided an opportunity for new properties to be rapidly spread.

But how did self-replicating organisms arise in the first place? It is fair to say that at the present time we simply do not know. No current hypothesis comes close to explaining how in the space of a mere 150 million years, the prebiotic environment that existed on planet Earth gave rise to life. That is not to say that reasonable hypotheses have not been put forward, but their statistical probability of accounting for the development of life still seems remote.

Fifty years ago, famous experiments by Stanley Miller and Harold Urey reconstructed a mixture of water and organic compounds that might have represented primeval circumstances on Earth. By applying an electrical discharge, these researchers were able to form small quantities of important biological building blocks, such as amino acids. The finding of small amounts of similar compounds within meteorites arriving from outer space has also been put forward as an argument that such complex organic molecules can arise from natural processes in the universe.

Beyond this point, however, the details become quite sketchy. How could a self-replicating information-carrying molecule assemble spontaneously from these compounds? DNA, with its phosphate-sugar backbone and intricately arranged organic bases, stacked neatly on top of one another and paired together at each rung of the twisted double helix, seems an utterly improbable molecule to have "just happened"—especially since DNA seems to possess no intrinsic means of copying itself. More recently, many investigators have pointed instead to RNA as the potential first life form, since RNA can carry information and in some instances it can also catalyze chemical reactions in ways that DNA cannot. DNA is something like the hard drive on your computer: it is supposed to be a stable medium in which to store information (though, as with your computer, bugs and snafus are always possible). RNA, by contrast, is more like a Zip disk or a flash drive—it travels around with its programming, and is capable of making things happen on its own. Despite substantial effort by multiple investigators, however, formation of the basic building blocks of RNA has not been achievable in a Miller-Urey type of experiment, nor has a fully self-replicating RNA been possible to design.

The profound difficulties in defining a convincing pathway for life's origin have led some scientists, most notably Francis Crick (who with James Watson discovered the DNA double helix), to propose that life forms must have arrived on Earth from outer space, either carried by small particles floating through interstellar space and captured by Earth's gravity or even brought here intentionally (or accidentally) by some ancient space traveler. While this might solve the dilemma of life's appearance on Earth, it does nothing to resolve the ultimate question of life's origin, since it simply forces that astounding event to another time and place even further back.

A word is in order here about an objection often raised by some critics to any possibility of the spontaneous origin of life on Earth, based on the Second Law of Thermodynamics. The Second Law states that in a closed system, where neither energy nor matter can enter or leave, the amount of disorder (more formally known as "entropy") will tend to increase over time. Since life forms are highly ordered, some have argued that it would therefore be impossible for life to have come into being without a supernatural creator. But this betrays a misunderstanding of the full meaning of the Second Law: order can certainly increase in some part of the system (as happens every day when you make the bed or put away the dishes), but that will require an input of energy, and the total amount of disorder in the entire system cannot decrease. In the case of the origin of life, the closed system is essentially the whole universe, energy is available from the sun, and so the local increase in order that would be represented by the first random assembly of macromolecules would in no way violate this law.

Given the inability of science thus far to explain the profound question of life's origins, some theists have identified the appearance of RNA and DNA as a possible opportunity for divine creative action. If God's intention in creating the universe was to lead to creatures with whom He might have fellowship, namely human beings, and if the complexity required to start the process of life was beyond the ability of the universe's chemicals to self-assemble, couldn't God have stepped in to initiate the process?

This could be an appealing hypothesis, given that no serious scientist would currently claim that a naturalistic explanation for the origin of life is at hand. But that is true today, and it may not be true tomorrow. A word of caution is needed when inserting specific divine action by God in this or any other area where scientific understanding is currently lacking. From solar eclipses in olden times to the movement of the planets in the Middle Ages, to the origins of life today, this "God of the gaps" approach has all too often done a disservice to religion (and by implication, to God, if that's possible). Faith that places God in the gaps of current understanding about the natural world may be headed for crisis if advances in science subsequently fill those gaps. Faced with incomplete understanding of the natural world, believers should be cautious about invoking the divine in areas of current mystery, lest they build an unnecessary theological argument that is doomed to later destruction. There are good reasons to believe in God, including the existence of mathematical principles and order in creation. They are positive reasons, based on knowledge, rather than default assumptions based on (a temporary) lack of knowledge.

In summary, while the question of the origin of life is a fascinating one, and the inability of modern science to develop a statistically probable mechanism is intriguing, this is not the place for a thoughtful person to wager his faith.

The Fossil Record

While amateur and professional scientists have been turning up fossils for centuries, these discoveries have reached a particularly intense phase over the last twenty years. Many of the previous gaps in understanding of the history of life on Earth are now being filled by the discovery of extinct species. Furthermore, their age can often be accurately estimated, based on the same process of radioactive decay that helped determine the age of the earth.

The vast majority of organisms that have ever lived on Earth have left absolutely no trace of their existence, since fossils arise only in highly unusual circumstances. (For example, a creature has to be caught in a certain type of mud or rock, without being picked apart by predators. Most bones rot and crumble. Most creatures decay.) Given that reality, it is actually rather amazing that we have such a wealth of information about organisms that have lived on this planet.

The timeline revealed by the fossil record is woefully incomplete, but still very useful. For example, only single-celled organisms appear in sediments that are older than about 550 million years, although it is possible that more complicated organisms existed prior to this time. Suddenly, approximately 550 million years ago, a great number of diverse invertebrate body plans appear in the fossil record. This is often referred to as the "Cambrian explosion," and is chronicled in highly readable form by the late Stephen Jay Gould, the most passionate and lyrical writer on evolution of his generation, in his book *Wonderful Life*. Gould himself questioned how evolution could account for the remarkable diversity of body plans that appeared in such a short span of time. (Other experts have been much less impressed with the claim that the Cambrian represents a discontinuity in life's complexity, though their writings have been less widely distributed to the general public. The so-called Cambrian explosion might, for example, reflect a change in conditions that allowed fossilization of a large number of species that had actually been in existence for millions of years.)

While attempts have been made by certain theists to argue that the Cambrian explosion is evidence of the intervention of some supernatural force, a careful examination of the facts does not seem to warrant this. This is another "God of the gaps" argument, and once again believers would be unwise to hang their faith upon such a hypothesis.

Current evidence suggests that the land remained barren until about 400 million years ago, at which point plants appeared on dry land, derived from aquatic life forms. A scarce 30 million years later, animals had also moved onto land. At one time, this step pointed to another gap: there appeared to be few transitional forms between sea creatures and land-dwelling tetrapods in the fossil record. Recent discoveries, however, have documented compelling examples of just this kind of transition.[3]

Beginning about 230 million years ago, dinosaurs dominated the earth. There is now general acceptance that their reign came to a sudden and catastrophic end approximately 65 million years ago, at the time of the collision of planet Earth with a large asteroid that fell in the general vicinity of what is now the Yucatan peninsula. Fine ash kicked up by this horrendous collision has been identified around the world, and the catastrophic climate changes that occurred from this vast amount of dust in the atmosphere apparently were too much for the dominant dinosaur species, leading to their demise and the subsequent rise of mammals.

That ancient asteroid collision is a tantalizing event. It may have been the only possible means by which the dinosaurs could have become extinct and mammals could have flourished. We probably wouldn't be here if that asteroid had not hit Mexico.

Most of us have a particular interest in the fossil record of humans, and here too the discoveries of the last few decades have been profoundly revealing. Bones of more than a dozen different hominid species have been discovered in Africa, with steadily increasing cranial capacity. The first specimens we recognize as modern *Homo sapiens* date from about 195,000 years ago. Other branches of hominid development appear to have encountered dead ends: the Neanderthals that existed in Europe until 30,000 years ago, and the recently discovered "hobbits," tiny people with small brains who lived on the island of Flores in Indonesia until extinction as recently as 13,000 years ago.

While there are many imperfections of the fossil record, and many puzzles remain to be solved, virtually all of the findings are consistent with the concept of a tree of life of related organisms. Good evidence exists for transitional forms from reptiles to birds, and from reptiles to mammals. Arguments that this model cannot explain certain species, such as whales, have generally fallen by the wayside as further investigation has revealed the existence of transitional species, often at precisely the date and place that evolutionary theory would predict.

Darwin's Revolutionary Idea

Born in 1809, Charles Darwin initially studied to become a cleric of the Church of England, but developed a deep interest in naturalism. Though the young Darwin was initially compelled by Paley's watchmaker argument, and saw design in nature as proof of a divine source, his views began to change when he traveled on the HMS *Beagle* from 1831 to 1836. He visited South America and the Galapagos Islands, where he examined the fossilized remains of ancient organisms and observed the diversity of life forms in isolated environments.

Building on these observations, and based on additional work over more than twenty years, Darwin developed the theory of evolution by natural selection. In 1859, faced with the possibility of being scooped by Alfred Russel Wallace, he finally wrote and published his ideas in the profoundly influential book *The Origin of Species*. Recognizing that the arguments in this book were likely to have broad reverberations, Darwin modestly commented near the end of the book, "When the views advanced by me in this volume, and by Mr. Wallace, or when analogous views on the origin of species are generally admitted, we can dimly foresee that there will be a considerable revolution in natural history."[4]

Darwin proposed that all living species are descended from a small set of common ancestors—perhaps just one. He held that variation within a species occurs randomly, and that the survival or extinction of each organism depends upon its ability to adapt to the environment. This he termed natural selection. Recognizing the potentially explosive nature of the argument, he hinted that this same process might apply to humankind, and developed this more fully in a subsequent book, *The Descent of Man*.

The Origin of Species engendered immediate and intense

controversy, though the reaction from religious authorities was not as universally negative as is often portrayed today. In fact, the notable conservative Protestant theologian Benjamin Warfield of Princeton accepted evolution as "a theory of the method of the divine providence,"[5] while arguing that evolution itself must have had a supernatural author.

There are many myths about public reaction to Darwin. For example, though there was a famous debate between Thomas H. Huxley (an ardent promoter of evolution) and Bishop Samuel Wilberforce, Huxley probably did not say (as legend has it) that he was unashamed to have a monkey for an ancestor, and would only be ashamed to be related to anyone who obscured the truth. Furthermore, far from his being ostracized by the religious community, Darwin was buried in Westminster Abbey.

Darwin himself was deeply concerned about the effect of his theory on religious belief, though in *The Origin of Species* he took pains to point out a possible harmonious interpretation: "I see no good reason why the views given in this volume should shock the religious feelings of anyone. . . . A celebrated author and divine has written to me that he 'has gradually learned to see that it is just as noble a conception of the deity to believe that he created a few original forms capable of self-development into other and needful forms, as to believe that he required a fresh act of creation to supply the voids caused by the action of his laws.' "[6]

Darwin even concludes *The Origin of Species* with the following sentence: "There is grandeur in this view of life, with its several powers, having been originally breathed by the Creator into a few forms or into one; and that, whilst this planet has gone cycling on according to the fixed law of gravity, from so simple a beginning, endless forms most beautiful and most wonderful have been, and are being evolved."[7]

Darwin's own personal beliefs remain ambiguous and seemed to vary throughout the last years of his life. At one time he said, "Agnostic would be the most correct description of my state of mind." At another time he wrote that he was greatly challenged by "the extreme difficulty, or rather the impossibility, of conceiving this immense and wonderful universe, including man with his capacity for looking far backwards and far into futurity, as the result of blind chance or necessity. When thus reflecting I feel compelled to look to a First Cause having an intelligent mind in some degree analogous to that of man; and I deserve to be called a Theist."[8]

No serious biologist today doubts the theory of evolution to explain the marvelous complexity and diversity of life. In fact, the relatedness of all species through the mechanism of evolution is such a profound foundation for the understanding of all biology that it is difficult to imagine how one would study life without it. Yet what area of scientific inquiry has generated more friction with religious perspectives than Darwin's revolutionary insight? From the circuslike Scopes "monkey trial" in 1925 right through to today's debates in the United States about the teaching of evolution in the schools, this battle shows no signs of ending.

DNA, THE HEREDITARY MATERIAL

Darwin's insight was all the more remarkable at the time because it lacked a physical basis. It took a century of work to discover just *how* there could be modifications in life's instruction book, in order to accommodate Darwin's "descent with modifications" idea.

Gregor Mendel, a relatively obscure Augustinian monk in what is now the Czech Republic, was a contemporary of Darwin and had read *The Origin of Species,* but they probably never met. Mendel was the first to demonstrate that inheritance could come in discrete packets of information. Through painstaking experiments with pea plants in the garden of his monastery, he concluded that hereditary factors involved in such attributes as the wrinkled or smooth appearance of peas were controlled by mathematical rules. He didn't know what a gene was, but his observations suggested that something like genes must exist.

Mendel's work was largely ignored for thirty-five years. Then, in one of the remarkable coincidences that occasionally arise in the history of science, it was rediscovered simultaneously by three other scientists within a few months of the turn of the twentieth century. In his famous studies on "inborn errors of metabolism," rare diseases that occurred in certain families in his medical practice, Archibald Garrod was able to show conclusively that Mendel's rules applied to humans, and that these disorders came about as a consequence of the same kind of inheritance that Mendel had appreciated in plants.

Mendel and Garrod added mathematical specificity to the notion of heritability in humans, though of course the reality of inherited characteristics such as skin and eye color was already familiar to anyone who was a close observer of our species. The mechanism behind these patterns remained obscure, however, as no one had successfully deduced the chemical basis of inheritance. Most researchers in the first half of the twentieth century assumed that inherited traits must be conveyed by proteins, since they appeared to be the most diverse molecules of living things.

It was not until 1944 that the microbiological experiments of Oswald T. Avery, Colin M. MacLeod, and Maclyn McCarty showed that it was DNA, not protein, that was capable of transferring inherited characteristics. Though the existence of DNA had been known for almost a hundred years, it was previously considered to be little more than nuclear packing material, of no particular interest.

Less than a decade later, a truly beautiful and elegant answer to the chemical nature of inheritance emerged. The furious race to determine the structure of DNA was won in 1953 by James Watson and Francis Crick, as is chronicled in Watson's entertaining book *The Double Helix*. Watson, Crick, and Maurice Wilkins, utilizing data produced by Rosalind Franklin, were able to deduce that the DNA molecule has the form of a double helix, a twisted ladder, and that its information-carrying capacity is determined by the series of chemical compounds that comprise the rungs of the ladder.

As a chemist, knowing how extraordinary the qualities of DNA really are, and how brilliant its solution is to the problem of coding life's design, I am in awe of this molecule. Let me try to explain just how elegant DNA really is.

As shown in Figure 4.1, the DNA molecule has a number of remarkable features. The outside backbone is made up of a monotonous ribbon of phosphates and sugars, but the interesting stuff lies on the inside. The rungs of the ladder are made up of combinations of four chemical components, called "bases." Let's call them (from the actual chemical names of these DNA bases) A, C, G, and T. Each of these chemical bases has a particular shape.

Now imagine that out of these four shapes, the A shape can fit neatly only on a ladder rung next to the T shape, and the G shape can fit only next to the C shape. These are "base pairs." Then you can picture the DNA molecule as a twisting ladder, with each rung made up of one base pair. There are four possible rungs: A-T, T-A, C-G, and G-C. If any single base is damaged on any one strand, it can be easily repaired by reference to the other strand: the only possible replacement for a T (for example) is another T. Perhaps most elegantly, the double helix immediately suggests a means of its self-copying, since each strand can be used as a template for the production of a new one. If you split all the pairs in half, cutting your ladder down the center of each rung, each half-ladder contains all the information needed to rebuild a complete copy of the original.

As a first approximation, one can therefore think of DNA as an instructional script, a software program, sitting in the nucleus of the cell. Its coding language has only four letters (or two bits, in computer terms) in its alphabet. A particular instruction, known as a gene, is made up of hundreds or thou-

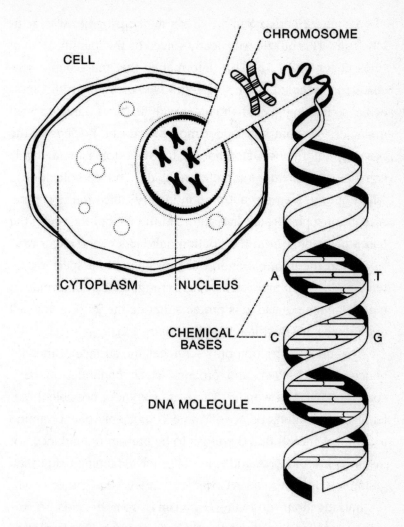

Figure 4.1. The double helix of DNA. Information is carried by the order of the chemical bases (A, C, G, and T). DNA is packaged into chromosomes, which reside in the nucleus of every cell.

sands of letters of code. All of the elaborate functions of the cell, even in as complex an organism as ourselves, have to be directed by the order of letters in this script.

At first, scientists had no idea how the program was actually "run." This puzzle was neatly solved by the identification of "messenger RNA." The DNA information that makes up a specific gene is copied into a single-stranded messenger RNA molecule, something like a half ladder with its rungs dangling from a single side. That half ladder moves from the nucleus of the cell (the information storehouse) to the cytoplasm (a highly complex gel mixture of proteins, lipids, and carbohydrates), where it enters an elegant protein factory called the ribosome. A team of sophisticated translators in the factory then read the bases protruding from the floating half-ladder messenger RNA to convert the information in this molecule into a specific protein, made up of amino acids. Three "rungs" of RNA information make one amino acid. It is proteins that do the work of the cell and provide its structural integrity (Figure 4.2).

This brief description only scratches the surface of the elegance of DNA, RNA, and protein, which continues to be a source of awe and wonder. There are sixty-four possible three-letter combinations of A, C, T, and G, but only twenty amino acids. That means that there has to be built-in redundancy: for instance, GAA in DNA and RNA codes for the amino acid called glutamic acid, but so does GAG.

Investigations of many organisms, from bacteria to humans, revealed that this "genetic code," by which information in DNA and RNA is translated into protein, is universal in all known organisms. No tower of Babel was to be allowed in the language of life. GAG means glutamic acid in the language of soil bacteria, the mustard weed, the alligator, and your aunt Gertrude.

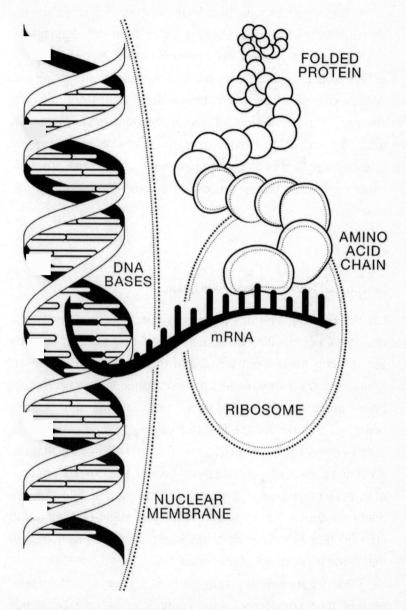

Figure 4.2. The flow of information in molecular biology: DNA → RNA → protein.

These advances gave birth to the field of molecular biology. The discovery of various other miniature chemical wonders, including proteins that act like scissors or glue, has enabled scientists to manipulate DNA and RNA by stitching together bits and pieces of these instructional molecules from different sources. This collection of molecular biological laboratory tricks, collectively referred to as recombinant DNA, has inspired a whole new field of biotechnology, and together with other advances promises to revolutionize the treatment of many diseases.

BIOLOGICAL TRUTH AND ITS CONSEQUENCES

For the believer who has previously taken the argument from design as a compelling demonstration of God's role in creating life, the conclusions put forward in this chapter can be unsettling. No doubt many readers have reasoned for themselves, or been taught in various religious settings, that the glorious beauty of a flower or the flight of an eagle could come about only as the consequence of a supernatural intelligence that appreciated complexity, diversity, and beauty. But now that molecular mechanisms, genetic pathways, and natural selection are being put forward to explain all this, you might be tempted to cry out, "Enough! Your naturalistic explanations are taking all the divine mystery out of the world!"

Do not fear, there is plenty of divine mystery left. Many people who have considered all the scientific and spiritual evidence still see God's creative and guiding hand at work. For me, there is not a shred of disappointment or disillusionment in these discoveries about the nature of life—quite the contrary! How marvelous and intricate life turns out to be! How deeply satisfying is the digital elegance of DNA! How aesthetically appealing and artistically sublime are the components of living things, from the ribosome that translates RNA into protein, to the metamorphosis of the caterpillar into the butterfly, to the fabulous plumage of the peacock attracting his mate! Evolution, as a mechanism, can be and must be true. But that says nothing about the nature of its author. For those who believe in God, there are reasons now to be more in awe, not less.

CHAPTER FIVE

Deciphering God's Instruction Book
The Lessons of the Human Genome

WHEN I WAS A RESEARCH FELLOW IN GENETICS at Yale in the early 1980s, the determination of the actual sequence of several hundred letters of the DNA code was an arduous undertaking. The methods were finicky, requiring many preparative steps, the use of expensive and dangerous reagents such as radioactive chemicals, and the manual pouring of ultrathin gels that were nearly always plagued with bubbles and other imperfections. The details are unimportant; the point is, it took forever, with lots of trial and error, just to sort out a few hundred letters of the human DNA code.

Despite those challenges, my first published paper on human genetics was based on DNA sequencing. I was studying the production of just one protein, found in the red blood cells

109

of the human fetus in utero, that is supposed to gradually disappear after birth, when babies begin to breathe with their own lungs. The protein is called fetal hemoglobin. Hemoglobin is the protein that allows red blood cells to deliver oxygen from our lungs to all of the rest of the body. Humans and some apes use a special version of hemoglobin before birth that helps extract oxygen from the mother's blood to nourish the growing fetus. During the first year of life, this fetal hemoglobin gradually turns off, and the adult form is produced instead. However, in a Jamaican family I was studying, substantial quantities of fetal hemoglobin continued to appear in adulthood. The cause of this "hereditary persistence of fetal hemoglobin" was of intense interest, because if we could learn how to trigger it on purpose in anyone, it could greatly reduce the ravages of sickle-cell anemia. Even 20 percent of fetal hemoglobin in the red blood cells of someone with sickle cell disease would essentially eliminate the painful crises and progressive organ damage.

I will never forget the day when my sequencing effort revealed a G instead of a C in a specific position just "upstream" of one of the genes that triggered fetal hemoglobin production. This single letter alteration turned out to be responsible for leaving the fetal program switched on in adults. I was thrilled but exhausted—it had taken eighteen months to discover this single altered letter of the human DNA code.

It was with some astonishment, then, that I learned three years later that a few visionary scientists had begun to discuss the possibility of determining the DNA sequence of the entire human genome, estimated to be about 3 billion base pairs in length. Surely this was not a goal that would be achieved during my lifetime.

We knew relatively little then about what the genome might contain. No one had actually seen the chemical bases of an individual human gene under the microscope (they were too tiny). Only a few hundred genes had been characterized, and estimates of how many genes the genome might contain varied wildly. Even the definition of a gene was (and is) in a bit of disarray—simple definitions that a gene constitutes a stretch of DNA that codes for a particular protein had been shaken by the discovery that the protein-coding regions of genes are interrupted by intervening DNA segments called introns. Depending on how the coding regions are subsequently spliced together in the RNA copy, one gene could sometimes code for several different (but related) proteins. Furthermore, there were long stretches of DNA in between genes that didn't seem to be doing very much; some even referred to these as "junk DNA," though a certain amount of hubris was required for anyone to call any part of the genome "junk," given our level of ignorance.

Despite all these uncertainties, there was no question how valuable a complete genome sequence would be. Hiding in this vast instruction book would be the parts list for human biology, as well as clues to a long list of diseases that we understand poorly and treat ineffectively. For me as a physician, the possibility of laying open the pages of this most powerful textbook of medicine was extremely compelling. And so, still junior in the ranks of academia, and uncertain of the practical realities of such an audacious plan, I joined the debate on the side of undertaking an organized program to sequence the human genome—which soon became known as the Human Genome Project.

My desire to see the human genome fully unveiled intensified considerably over the next few years. Leading a fledgling research laboratory of earnest and hardworking graduate students and postdoctoral fellows, I had decided to pursue the genetic basis of certain diseases that had so far resisted all attempts at discovery. Foremost among these was cystic fibrosis (CF), the most common potentially fatal genetic disorder of northern Europeans. The disease is usually diagnosed in an infant or young child who fails to gain weight and suffers from repeated respiratory infections. Picking up on information from observant mothers who had noted that these children taste salty when kissed, physicians had identified a high concentration of chloride in a child's sweat as the diagnostic hallmark. We also knew that CF patients had thick, sticky secretions in their lungs and pancreas—but we had no real clue as to the likely function of the gene that must have gone awry.

I first encountered CF when serving as a medical intern in the late 1970s. In the 1950s kids with this disease had rarely survived to age ten. Steady improvements in managing the symptoms—replacing enzymes in the pancreas, treating lung infections with better antibiotics, and improving nutrition and physical therapy—had gradually extended the life span of CF patients so that many of them in the 1970s were surviving to attend college, marry, and enter the workforce. But the long-term prospects for cure were still bleak. Without a fundamental understanding of the genetic defect, medical researchers were just feeling their way in the dark. All we knew was that somewhere in the 3 billion letters of the DNA code, at least one letter had gone wrong in a vulnerable location.

To find such a subtle misspelling seemed an almost insurmountable problem of scale. But the other thing we knew about CF is that it is inherited in a recessive pattern. To understand the meaning of this, it is important to consider that we all have two copies of each gene, one inherited from Mom and one from Dad. (The exceptions are genes on the X and the Y chromosomes, which are present in only a single copy in males.) In a recessive disease like CF, a child is affected only if *both* copies of the gene are faulty. For that to happen, both parents must carry a flawed copy—but since individuals with one normal and one faulty copy are entirely well, these carriers are generally unaware of their status (about one in thirty individuals of northern European ancestry is a CF carrier, and most of them have no family history of the disease).

The genetic basis of CF thus allowed an interesting exercise in DNA sleuthing: even without knowing anything about the responsible gene, researchers could track the inheritance of hundreds of random bits of DNA from all over the genome in CF families with multiple siblings, looking for DNA fragments that were predictive of which siblings had CF and which did not. Such fragments would by necessity have to be located close to the CF gene. We couldn't read all 3 billion pairs of letters, but we could randomly shine a flashlight on a few million here, a few million there, and look for any correlation with the disease. We had to do this hundreds and hundreds of times, but the genome is a bounded set of information—so if we kept at it, we were confident of locating the right neighborhood.

That was accomplished, to the astonishment and delight of scientists and families alike, in 1985—and demonstrated that the CF gene must reside somewhere within a 2 million base-pair segment of DNA on chromosome 7. But the hard part had really just begun. To employ an analogy I often used at that time to explain why this was such a hard problem, the search was like looking for a single burned-out lightbulb in the basement of a house somewhere in the United States. The family studies were a powerful beginning, in the sense that they allowed us to identify the right state and ultimately the right county. But this was the view from twenty thousand feet, and this strategy could take us no further. A house-to-house search, lightbulb by lightbulb, was required.

We didn't even have a map of the territory. This part of chromosome 7, like most of the genome, had never been explored in 1985. To pursue the metaphor, there were no street maps of towns and villages, no blueprints of buildings, certainly no inventory of lightbulbs. The work was brutal.

My team and I had invented a method called "chromosome jumping," which allowed us to move across our 2 million base-pair target in pogo-stick leaps, rather than crawling along in the traditional way. That helped by enabling the house-to-house searches to be initiated in multiple locations at once. But the challenge was still almost overwhelming, and many in the scientific community thought that this approach was so impractical that it would just never work for a human disease. In 1987, faced with limited resources and mounting frustration, my lab joined forces with that of Lap-Chee Tsui, a talented Ph.D. researcher at the Hospital for Sick Children in

Toronto. Our merged labs pressed on with renewed energy. The search was like a detective story—we knew the mystery would eventually be solved on the last page, but we didn't know how long it would take to get there. There were clues and blind alleys aplenty. After getting excited for the third or fourth time about a possible answer, only to have it collapse the next day because of new data, we stopped allowing ourselves to be very optimistic about anything. We found it hard to keep explaining to colleagues why we hadn't found the gene yet, or alternatively why we hadn't just given up. At one point, seeking another metaphor to explain the difficulty of the problem, I even went to a local Michigan farm to have my picture taken holding a sewing needle while sitting atop a large haystack.

But one rainy night in May 1989, the answer finally came. There, spilling out of the fax machine Lap-Chee and I had set up in the Yale dormitory where we were both attending a meeting, was the data from that day's work in the lab—showing unequivocally that a deletion of just three letters of the DNA code (CTT, to be exact) in the protein-coding part of a previously unknown gene was the cause of cystic fibrosis in the majority of patients. Soon after, we and others were able to show that this mutation and other less common misspellings in this same gene, now called CFTR, account for virtually all cases of the disease.

There it was—the proof that we could actually find that burned-out lightbulb, that we could identify a disease gene by progressively narrowing its chromosomal position. It was a grand moment of celebration. The road had been long and hard, but now hopes were high that research on finding a cure could get under way in earnest.

At a subsequent gathering of thousands of CF researchers, families, and clinicians, I wrote a song to commemorate the gene discovery. Music has always helped me to express and experience things in ways that simple words cannot. Though my guitar skills are only modest, I find great joy in those moments where people raise their voices together. That experience is made up more of spirit than of science. I found myself unable to hold back the tears as these legions of good people rose from their seats and sang along with the chorus:

> Dare to dream, dare to dream,
> All our brothers and sisters breathing free.
> Unafraid, our hearts unswayed,
> Till the story of CF is history.

The next steps proved harder than expected, and the story of CF is regrettably still not history. But the gene finding was indeed gratifying, and started CF research on a course toward what we all expect will be ultimate victory. Adding up all of the work that had been done by the more than two dozen teams worldwide to find the CF gene, it had taken ten years and more than $50 million to identify this one gene for this one disease. And CF was supposed to be one of the easiest—since it was a relatively common disease that followed Mendel's rules of inheritance precisely. How could we ever imagine extending this work to the hundreds of rarer genetic diseases that urgently needed unraveling? Even more challenging, how could we imagine applying this same strategy to diseases like diabetes, schizophrenia, heart disease, or the common cancers, where we know hereditary factors are critically important but the best evidence indicates that many different genes are involved, and no single gene contributes a very strong effect? In those instances, there might be a dozen or more lightbulbs to discover, and they weren't even expected to be burned out—just subtly dimmer than they should be. If there was to be any hope of succeeding in these more difficult circumstances, we simply had to have detailed and accurate information about every nook and cranny in the human genome. We needed a house-by-house map of the entire country.

Arguments about the wisdom of the project raged furiously during the late 1980s.[1] While most scientists had to agree that the information would eventually be useful, the sheer magnitude of the project made it seem almost unattainable. Furthermore, it was already clear that only a small fraction of the genome was dedicated to coding for protein, and the wisdom of sequencing the rest (the "junk DNA") was debatable. One well-known scientist wrote: "Sequencing the genome would be about as useful as translating the complete works of Shakespeare into cuneiform, but not quite as feasible or as easy to interpret."

Another wrote: "It makes no sense . . . geneticists would be wading through a sea of drivel to emerge dry shod on a few tiny islands of information." Much of the concern was really based, however, on the potential cost of such an enterprise, and the possibility that it might siphon funds away from the rest of the biomedical research enterprise. The best antidote for that concern was to expand the pie, and find new funds for the project. That was neatly engineered in the United States by the new director of the genome project, none other than Jim Watson himself, the codiscoverer of the DNA double helix. Watson, at that time the unrivaled rock star of biology, convinced Congress to take a risk on this new endeavor.

Jim Watson ably oversaw the first two years of the U.S. Human Genome Project, establishing genome centers and recruiting some of the best and brightest scientists of the current generation to work on the project. Much skepticism remained, however, about whether the project would be able to deliver on its fifteen-year timetable, given that many of the technologies needed for accomplishing the goals had not even been invented yet. In 1992, a crisis occurred when Watson suddenly left the project after a public argument with the director of the National Institutes of Health about the wisdom of patenting bits and pieces of DNA (to which Watson was strongly opposed).

An intense national search ensued to find a new director. No one was more surprised than I to find the selection process converging on me. Being quite happy at the time leading a genome center at the University of Michigan, and never having imagined myself as a federal employee, I initially indicated no interest. But the decision haunted me. There was only one Human Genome Project. This was going to be done only once in human history. If it succeeded, the consequences for medicine would be unprecedented. As a believer in God, was this one of those moments where I was somehow being called to take on a larger role in a project that would have profound consequences for our understanding of ourselves? Here was a chance to read the language of God, to determine the intimate details of how humans had come to be. Could I walk away? I have always been suspicious of those who claim to perceive God's will in moments such as this, but the awesome significance of this adventure, and the potential consequences for humankind's relationship with the Creator, could hardly be ignored.

Visiting my daughter in North Carolina in November 1992, I spent a long afternoon praying in a little chapel, seeking guidance about this decision. I did not "hear" God speak—in fact, I have never had that experience. But during those hours, ending in an evensong service that I had not expected, a peace settled over me. A few days later, I accepted the offer.

The next ten years were a wild roller coaster of experiences. The original goals of the Human Genome Project were incredibly ambitious, but we set aggressive milestones and held ourselves accountable for achieving them. There were moments of great frustration, when methods that seemed very promising in initial tests turned out to fail spectacularly on a larger scale. Friction sometimes broke out among members of our scientific team, and it was my job to serve as mediator. Some centers failed to keep up the pace and had to be phased out, much to the dismay of their leaders. But there were also moments of triumph, as challenging goals were met and new medical insights began to pile up. By 1996, we were ready to start piloting the actual large-scale sequencing of the human genome, using a process that was vastly more technically advanced and cost effective than it had been in 1985 during my hunt for the CF gene. In a defining moment, those of us leading the international public project made immediate access to the data a requirement for participation, and agreed that no patents of any sort would be filed on the DNA sequence. We could not justify even a single day passing where researchers around the world, aiming to understand important medical problems, would not have free and open access to the data being produced.

The next three years proved fruitful, and by 1999 we were ready to accelerate dramatically. But a new challenge appeared on the horizon. Sequencing the entire human genome had previously been considered unattractive as a commercial enterprise, but as the value of the information became more and more apparent, and the costs of sequencing came down, a major challenge to the public Human Genome Project was mounted by a private company. Craig Venter, the leader of the company soon to be named Celera, announced that he would carry out large-scale sequencing on the human genome, but would file patents on many of the genes, and would keep the data in a subscription database that would require significant payment for access.

The idea that the human-genome sequence might become private property was deeply distressing. Even more of concern, questions began to be raised in the Congress about whether it made sense to continue to spend taxpayers' money on a project that might better be carried out in the private sector—though no actual data from the Celera team was available, and the scientific strategy that Venter aimed to pursue was unlikely to yield a truly finished and highly accurate sequence. Yet a constant stream of claims of higher efficiency poured out of the well-oiled Celera public relations machine, which also sought to label the public project as slow and bureaucratic. Given that the work of the Human Genome Project was being done in some of the world's finest universities by some of the most creative and dedicated scientists on the planet, that was a little hard to take. Yet the press loved the controversy. Many articles were written about "the race" to sequence the genome, and about Venter's yacht and my motorcycle. What drivel! What most observers seemed to miss was that this was not, at its core, a debate about who would do the work faster or cheaper (both Celera and the public project were now well situated to deliver on this). It was instead a battle of ideals—would the human genome sequence, our shared inheritance, become a commercial commodity, or a universal public good?

No effort could now be spared by our team. Our twenty public genome centers in six countries ran around the clock. In the space of just eighteen months, after generating a thousand base pairs a second, seven days a week, twenty-four hours a day, a draft covering 90 percent of the human genome sequence was in hand. All of the data continued to be released every twenty-four hours. For their part, Celera also generated large amounts of data, but it remained out of view in their private database. Recognizing that they could also take advantage of public data, Celera ultimately stopped at only half the production they had planned. Ultimately more than half of the Celera genome assembly turned out to consist of public data.

The attention to "the race" was becoming unseemly, and threatened to diminish the importance of the goal. In late

April 2000, with both Celera and the public project poised to announce that a draft had been achieved, I approached a mutual friend of Venter and myself (Ari Patrinos of the Department of Energy's genome program), and asked him to set up a secret meeting. Over beer and pizza in Ari's basement, Venter and I worked out a plan for a simultaneous announcement.

Thus, as described in the opening pages of this book, I found myself standing next to the president of the United States in the East Room of the White House on June 26, 2000, announcing that a first draft of the human instruction book had been determined. The language of God was revealed.

Over the next three years, I had the privilege of continuing to lead the public project to refine this draft sequence, closing the remaining gaps, pushing the accuracy of the information to a very high level, and continuing to deposit all of the data into public databases on a daily basis. In April 2003, in the month that marked the fiftieth anniversary of Watson and Crick's publication on the double helix, we announced the completion of all of the goals of the Human Genome Project. As the project manager of the enterprise, I was intensely proud of the more than two thousand scientists who had accomplished this remarkable feat, one that I believe will be seen a thousand years from now as one of the major achievements of humankind.

At a subsequent celebration of the success of the Human Genome Project, sponsored by the Genetic Alliance, a heartwarming organization that exists to encourage and empower families who face rare genetic diseases, I rewrote the familiar folk song "All the Good People" to fit the occasion. All joined in the chorus:

> This is a song for all the good people,
> All the good people who are part of this family.
> This is a song for all the good people,
> We're joined together by this common thread.

I wrote another verse, about what many of those families were going through as they struggled to cope with rare diseases in themselves or their children:

> This is a song for those who are suffering,
> Your strength and your spirit have touched
> one and all.
> It's your dedication that's our inspiration,
> Because of your courage, you help us stand tall.

And finally, I added a verse about the genome:

> It's a book of instructions, a record of history,
> A medical textbook, it's all these entwined
> It's of the people, by the people,
> It's for the people, it's yours and it's mine.

For me, as a believer, the uncovering of the human genome sequence held additional significance. This book was written in the DNA language by which God spoke life into being. I felt an overwhelming sense of awe in surveying this most significant of all biological texts. Yes, it is written in a language we understand very poorly, and it will take decades, if not centuries, to understand its instructions, but we had crossed a one-way bridge into profoundly new territory.

SURPRISES FROM THE FIRST READING OF THE GENOME

Entire books have been written about the Human Genome Project (probably too many, in fact).[2] Perhaps I'll write my own someday, hopefully with sufficient hindsight to avoid some of the breathless pronouncements of many of the currently popular depictions. It is not the purpose of this book, however, to dwell further upon that remarkable experience, but rather to reflect upon the ways that a modern understanding of science can be harmonized with a belief in God.

In that regard, it is interesting to look carefully at the genome of humankind, and to compare it with the genomes of many other organisms that have now been sequenced. When we survey the vast expanse of the human genome, 3.1 billion letters of the DNA code arrayed across twenty-four chromosomes, several surprises are immediately apparent.

One surprise is just how little of the genome is actually used to code for protein. Though limitations of both our experimental and computational methods still prevent a precise estimate, there are only about 20,000–25,000 protein-coding genes in the human genome. The total amount of DNA used by those genes to code for protein adds up to a measly 1.5 percent of the total. After a decade of expecting to find at least 100,000 genes, many of us were stunned to discover that God writes such short stories about humankind. That was especially shocking in the context of the fact that the gene counts for other simpler organisms such as worms, flies, and simple plants seem to be in about the same range, namely around 20,000.

Some observers have taken this as a real insult to human complexity. Have we been deluding ourselves about our special place in the animal kingdom? Well, not really—clearly gene count must not be the whole story. By any estimation, the biological complexity of human beings considerably exceeds that of a roundworm, with its total of 959 cells, even though the gene count is similar for both. And certainly no other organism has sequenced its own genome! Our complexity must arise not from the number of separate instruction packets, but from the way they are utilized. Perhaps our component parts have learned how to multitask?

Another way to think about this is to consider the metaphor of language. The average educated English speaker has a vocabulary of about 20,000 words. Those words can be used to construct rather simple documents (such as an owner's manual for your car) or much more complex works of literature such as James Joyce's *Ulysses*. In the same way, worms, insects, fish, and birds apparently need an extensive vocabulary of 20,000 genes to function, though they use these resources in less elaborate ways than we do.

Another striking feature of the human genome comes from the comparison of different members of our own species. At the DNA level, we are all 99.9 percent identical. That similarity applies regardless of which two individuals from around the world you choose to compare. Thus, by DNA analysis, we humans are truly part of one family. This remarkably low genetic diversity distinguishes us from most other species on the planet, where the amount of DNA diversity is ten or sometimes even fifty times greater than our own. An alien visitor sent here to examine life forms on earth might have many interesting things to say about humankind, but most certainly he would comment on the surprisingly low level of genetic diversity within our species.

Population geneticists, whose discipline involves the use of mathematical tools to reconstruct the history of populations of animals, plants, or bacteria, look at these facts about the human genome and conclude that they point to all members of our species having descended from a common set of founders, approximately 10,000 in number, who lived about 100,000 to 150,000 years ago. This information fits well with the fossil record, which in turn places the location of those founding ancestors most likely in East Africa.

Another profoundly interesting consequence of the study of multiple genomes has been the ability to do detailed comparisons of our own DNA sequence with that of other organisms. Using a computer, one can pick a certain stretch of human DNA and assess whether there is a similar sequence in some other species. If one picks the coding region of a human gene (that is, the part that contains the instructions for a protein), and uses that for the search, there will nearly always be a highly significant match to the genomes of other mammals. Many genes will also show discernible but imperfect matches to fish. Some will even find matches to the genomes of simpler organisms such as fruit flies and roundworms. In some particularly striking examples, the similarity will extend all the way down to genes in yeast and even to bacteria.

If, on the other hand, one chooses a bit of human DNA that lies between genes, then the likelihood of being able to find a similar sequence in the genomes of other distantly related organisms decreases. It does not disappear entirely; with careful computer searching, about half of all such fragments can be aligned with other mammalian genomes, and almost all of them align nicely with the DNA of other nonhuman primates. Table 5.1 shows the percentages for success of this kind of matchup, divided up into various categories.

What does all this mean? At two different levels, it provides powerful support for Darwin's theory of evolution, that is,

	Gene Sequence That Codes for Protein	Random DNA Segment Between Genes
Chimpanzee	100%	98%
Dog	99%	52%
Mouse	99%	40%
Chicken	75%	4%
Fruit fly	60%	~0%
Roundworm	35%	~0%

Table 5.1 Likelihood of Finding a Similar DNA Sequence in the Genome of Other Organisms, Starting with a Human DNA Sequence

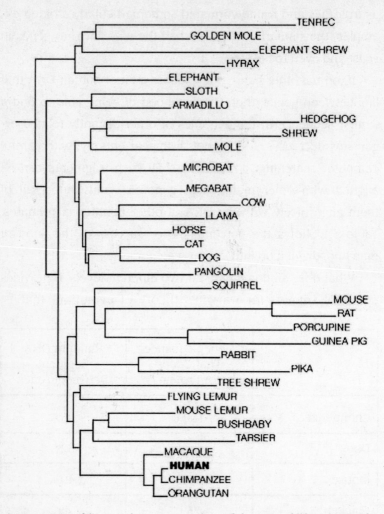

Figure 5.1 On this page is a current view of the tree of life, where relationships between different mammalian species are inferred solely by a comparison of their DNA sequences. The length of the branches represents the degree of difference between species—thus the DNA sequences of mouse and rat are more closely related than those of mouse and squirrel, and the DNA sequences of human and chimpanzee are more closely related than those of human and macaque. Opposite, for an interesting historical comparison, is a page from Darwin's 1837 notebook, where the words "I think" are followed by his own idea of the tree of life that connects different species.

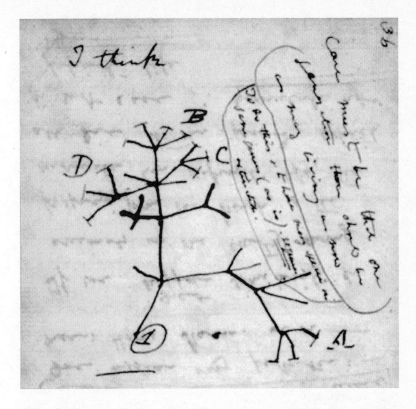

descent from a common ancestor with natural selection operating on randomly occurring variations. At the level of the genome as a whole, a computer can construct a tree of life based solely upon the similarities of the DNA sequences of multiple organisms. The result is shown in Figure 5.1. Bear in mind that this analysis does not utilize any information from the fossil record, or from anatomic observations of current life forms. Yet its similarity to conclusions drawn from studies of comparative anatomy, both of existent organisms and of fossilized remains, is striking. Second, within the genome, Darwin's theory predicts that mutations that do not affect function (namely, those located in "junk DNA") will accumulate steadily over time. Mutations in the coding region of genes, however, are expected to be observed less frequently, since most of these will be deleterious, and only a rare such event will provide a selective advantage and be retained during the evolutionary process. That is exactly what is observed. This latter phenomenon even applies to the fine details of the coding regions of genes. From the previous chapter, you may recall that the genetic code is degenerate: for example, GAA and GAG both code for glutamic acid. That means that it is possible for some mutations in the coding region to be "silent," where the encoded amino acid is not altered by the change, and so no penalty is paid. When comparing DNA sequences of related species, silent differences are much more common in the coding regions than those that alter an amino acid. That is exactly what Darwin's theory would predict. If, as some might argue, these genomes were created by individual acts of special creation, why would this particular feature appear?

Darwin and DNA

Charles Darwin was intensely insecure about his theory of evolution. Perhaps that accounts for the nearly twenty-five years that passed between his development of the idea and his publication of *The Origin of Species*. There must have been many times when Darwin wished that he could go back millions of years in time and actually observe all of the events that his theory predicted. Of course he couldn't do that, and we can't do that today either. But lacking a time machine, Darwin could hardly have imagined a more compelling digital demonstration of his theory than what we find by studying the DNA of multiple organisms.

In the mid-nineteenth century, Darwin had no way of knowing what the mechanism of evolution by natural selection might be. We can now see that the variation he postulated is supported by naturally occurring mutations in DNA. These are estimated to occur at a rate of about one error every 100 million base pairs per generation. (That means, by the way, that since we all have two genomes of 3 billion base pairs each, one from our mother and one from our father, we all have roughly sixty new mutations that were not present in either of our parents.)

Most of those mutations occur in parts of the genome that are not essential, and therefore they have little or no consequence. The ones that fall in the more vulnerable parts of the genome are generally harmful, and are thus rapidly culled out of the population because they reduce reproductive fitness. But on rare occasions, a mutation will arise by chance that offers a slight degree of selective advantage. That new DNA "spelling" will have a slightly higher likelihood of being passed on to future offspring. Over the course of a very long period of time, such favorable rare events can become widespread in all members of the species, ultimately resulting in major changes in biological function.

In some instances, scientists are even catching evolution in the act, now that we have the tools to track these events. Some critics of Darwinism like to argue that there is no evidence of "macroevolution" (that is, major change in species) in the fossil record, only of "microevolution" (incremental change within a species). We have seen finch beaks change shape over time, they argue, depending upon changing food sources, but we haven't seen new species arise.

This distinction is increasingly seen to be artificial. For example, a group at Stanford University is engaged in an intense effort to understand the wide diversity of body armor in stickleback fish. Sticklebacks that live in salt water typically have a continuous row of three dozen armor plates extending from head to tail, but freshwater populations from many different parts of the world, where predators are fewer, have lost most of these plates.

The freshwater sticklebacks apparently arrived in their current locations ten to twenty thousand years ago after widespread melting of glaciers at the end of the last Ice Age. A careful comparison of the genomes of the freshwater fish has identified a specific gene, EDA, whose variants have repeatedly and independently appeared in a freshwater situation, resulting in loss of the plates. Interestingly, humans also have an EDA gene, and spontaneous mutations in that gene result in defects in hair, teeth, sweat glands, and bone. It is not hard to see how the difference between freshwater and saltwater sticklebacks could be extended, to generate all kinds of fish. The distinction between macroevolution and microevolution is therefore seen to be rather arbitrary; larger changes that result in new species are a result of a succession of smaller incremental steps.

Evolution is also seen to be at work in everyday experience by the rapid variations in certain disease-causing viruses, bac-

teria, and parasites that can cause major public health out-
breaks. When I contracted malaria in West Africa in 1989, that
was despite having taken the recommended prophylaxis
(chloroquine). Randomly occurring natural variations in the
genome of the malarial parasite, subjected to selection over
many years of heavy use of chloroquine in that part of the
world, had ultimately resulted in a pathogen that was resistant
to the drug, and therefore spread rapidly. Similarly, rapid evolutionary changes in the HIV virus that causes AIDS have provided a major challenge for vaccine development, and are the
major cause of ultimate relapse in those treated with drugs
against AIDS. Even more in the public eye, the fears of a pan-
demic influenza outbreak from the H5N1 strain of avian flu are
based upon the high likelihood that the current strain, devastating as it already is to chickens and a few humans who have had
close contact with them, will evolve into a form that spreads
easily from person to person. Truly it can be said that not only
biology but medicine would be impossible to understand without the theory of evolution.

WHAT DOES THIS SAY ABOUT HUMAN EVOLUTION?

Applying evolutionary science to sticklebacks may be one thing,
but what about ourselves? Since Darwin's time, people of many
different worldviews have been particularly motivated to understand how revelations about biology and evolution apply to that
special class of animals, human beings.

The study of genomes leads inexorably to the conclusion

that we humans share a common ancestor with other living things. Some of that evidence is shown in Table 5.1, where the similarity between the genomes of ourselves and other organisms is displayed. This evidence alone does not, of course, prove a common ancestor; from a creationist perspective, such similarities could simply demonstrate that God used successful design principles over and over again. As we shall see, however, and as was foreshadowed above by the discussion of "silent" mutations in protein-coding regions, the detailed study of genomes has rendered that interpretation virtually untenable—not only about all other living things, but also about ourselves.

As a first example, let us look at a comparison of the human and mouse genomes, both of which have been determined at high accuracy. The overall size of the two genomes is roughly the same, and the inventory of protein-coding genes is remarkably similar. But other unmistakable signs of a common ancestor quickly appear when one looks at the details. For instance, the order of genes along the human and the mouse chromosomes is generally maintained over substantial stretches of DNA. Thus, if I find human genes A, B, and C in that order, I am likely to find that the mouse has counterparts of A, B, and C also placed in that same order, although the spacing between the genes may have varied a bit (Figure 5.2). In some instances, this correlation extends over substantial distances; virtually all of the genes on human chromosome 17, for instance, are found on mouse chromosome 11. While one might argue that the order of genes is critical in order for their function to occur properly, and therefore a designer might have

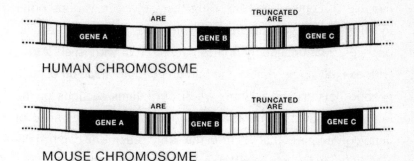

HUMAN CHROMOSOME

MOUSE CHROMOSOME

Figure 5.2 The order of genes along a chromosome is often the same in humans and mice, though the precise spacing between genes may vary somewhat. Thus, if you find the order of three genes to be A, B, and C along a human chromosome, you are very likely to find the mouse counterparts of the A, B, and C genes in the same order on the mouse chromosome. Furthermore, now that complete genome sequences of both humans and mice are available, it is possible to identify in the spaces between genes the remnants of many "jumping genes." These are transposable elements that can insert themselves at random into the genome, and even continue to do so at a low level today. By DNA sequence analysis, some of these elements have acquired many mutations compared with the original jumping gene, and thus appear to be very old; these are referred to as ancient repetitive elements (AREs). Interestingly, these ancient elements are often found in similar locations in the mouse and human genomes (as in this example, where an ARE is present between gene A and gene B in both human and mouse). Particularly interesting are examples where the ARE was truncated at a precise base pair at the time of insertion, losing part of its DNA sequence and all possibility of future function (as in the example between gene B and gene C). Finding a precisely truncated ARE in the same place in both human and mouse genomes is compelling evidence that this insertion event must have occurred in an ancestor that was common to both the human and the mouse.

maintained that order in multiple acts of special creation, there is no evidence from current understanding of molecular biology that this restriction would need to apply over such substantial chromosomal distances.

Even more compelling evidence for a common ancestor comes from the study of what are known as ancient repetitive elements (AREs). These arise from "jumping genes," which are capable of copying and inserting themselves in various other locations in the genome, usually without any functional consequences. Mammalian genomes are littered with such AREs, with roughly 45 percent of the human genome made up of such genetic flotsam and jetsam. When one aligns sections of the human and mouse genomes, anchored by the appearance of gene counterparts that occur in the same order, one can usually also identify AREs in approximately the same location in these two genomes (Figure 5.2).

Some of these may have been lost in one species or the other, but many of them remain in a position that is most consistent with their having arrived in the genome of a common mammalian ancestor, and having been carried along ever since. Of course, some might argue that these are actually functional elements placed there by the Creator for a good reason, and our discounting of them as "junk DNA" just betrays our current level of ignorance. And indeed, some small fraction of them may play important regulatory roles. But certain examples severely strain the credulity of that explanation. The process of transposition often damages the jumping gene. There are AREs throughout the human and mouse genomes that were truncated when they landed, removing any possibility of their functioning. In many instances, one can identify a decapitated and utterly defunct ARE in parallel positions in the human and the mouse genome (Figure 5.2).

Unless one is willing to take the position that God has placed these decapitated AREs in these precise positions to confuse and mislead us, the conclusion of a common ancestor

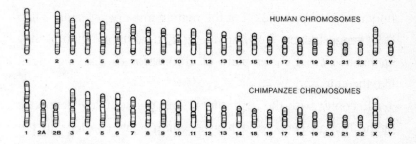

Figure 5.3 The human and chimpanzee chromosomes, or "karyotypes." Note the marked similarity in size and number, with a notable exception: human chromosome 2 seems to be made up of a head-to-head fusion of two intermediate-size chimp chromosomes (here labeled 2A and 2B).

for humans and mice is virtually inescapable. This kind of recent genome data thus presents an overwhelming challenge to those who hold to the idea that all species were created ex nihilo.

The placement of humans in the evolutionary tree of life is only further strengthened by a comparison with our closest living relative, the chimpanzee. The chimpanzee genome sequence has now been unveiled, and it reveals that humans and chimps are 96 percent identical at the DNA level.

A further example of this close relationship stems from examination of the anatomy of human and chimpanzee chromosomes. Chromosomes are the visible manifestation of the DNA genome, apparent in the light microscope at the time that a cell divides. Each chromosome contains hundreds of genes. Figure 5.3 shows a comparison of the chromosomes between a human and a chimpanzee. The human has twenty-three pairs of chromosomes, but the chimpanzee has twenty-four. The difference in the chromosome number appears to be a consequence of two ancestral chromosomes having fused together to generate human chromosome 2. That the human must be a fusion is further suggested by studying the gorilla and orangutan—they each have twenty-four pairs of chromosomes, looking much like the chimp.

Recently, with the determination of the complete sequence of the human genome, it has become possible to look at the precise location where this proposed chromosomal fusion must have happened. The sequence at that location—along the long arm of chromosome 2—is truly remarkable. Without getting into the technical details, let me just say that special sequences occur at the tips of all primate chromosomes. Those sequences generally do not occur elsewhere. But they are found right where evolution would have predicted, in the middle of our fused second chromosome. The fusion that occurred as we evolved from the apes has left its DNA imprint here. It is very difficult to understand this observation without postulating a common ancestor.

Yet another argument for the common ancestry of chimps and humans comes from the peculiar observation of what are called pseudogenes. Those are genes that have almost all of the properties of a functional DNA instruction packet, but are afflicted by one or more glitches that turn their script into gibberish. When one compares chimp and human, occasional genes appear that are clearly functional in one species but not in the other, because they have acquired one or more deleterious mutations. The human gene known as caspase-12, for instance, has sustained several knockout blows, though it is found in the identical relative location in the chimp. The chimp caspase-12 gene works just fine, as does the similar gene in nearly all mammals, including mice. If humans arose as a consequence of a supernatural act of special creation, why would God have gone to the trouble of inserting such a nonfunctional gene in this precise location?

We can also now begin to explain the origins of a tiny fraction of the more mechanical differences between us and our closest relatives, some of which may play crucial roles in our humanness. In one example, a gene for a jaw muscle protein (MYH16) appears to have mutated into a pseudogene in humans. It continues to play a significant role in the development and strength of the jaw muscles in other primates. It is just conceivable that the inactivation of this gene led to a reduction in the mass of the human jaw muscle. Most apes have relatively larger and stronger jaws than we do. Human and ape skulls must, among other things, serve as an anchor for these jaw muscles. It is possible that the development of weaker jaws paradoxically allowed our skulls to expand upward, and accommodate our larger brains. This is clearly speculation, of course, and other genetic changes would be necessary to account for the much larger brain cortex that represents a major component of the difference between humans and chimpanzees.

In another example, much interest has recently surrounded the gene called FOXP2 because of its potential role in the development of language. The story of FOXP2 began with the identification of a single family in England where members of three generations had severe difficulty in speaking. They struggled to process words according to grammatical rules, to understand complex sentence structure, and to move the muscles of their mouths, faces, and voice boxes, to articulate certain sounds.

In a tour de force of genetic sleuthing, the affected family members were found to have a single letter of the DNA code misspelled in the FOXP2 gene on chromosome 7. The fact that a single gene with a subtle misspelling could cause such profound language deficits, without other obvious consequences, was quite surprising.

The surprise rapidly escalated when it was shown that the sequence of this same FOXP2 gene has been remarkably stable in nearly all mammals. The most dramatic exception, however, is humans, where two significant changes have occurred in the coding region of the gene, apparently as recently as a hundred thousand years ago. The hypothesis suggested by these data is that these recent changes in FOXP2 may have in some way contributed to the development of language in human beings.

At this point, godless materialists might be cheering. If humans evolved strictly by mutation and natural selection, who needs God to explain us? To this, I reply: I do. The comparison of chimp and human sequences, interesting as it is, does not tell us what it means to be human. In my view, DNA sequence alone, even if accompanied by a vast trove of data on biological function, will never explain certain special human attributes, such as the knowledge of the Moral Law and the universal search for God. Freeing God from the burden of special acts of creation does not remove Him as the source of the things that make humanity special, and of the universe itself. It merely shows us something of how He operates.

EVOLUTION: A THEORY OR A FACT?

The examples reported here from the study of genomes, plus others that could fill hundreds of books of this length, provide the kind of molecular support for the theory of evolution that has convinced virtually all working biologists that Darwin's framework of variation and natural selection is unquestionably correct. In fact, for those like myself working in genetics, it is almost impossible to imagine correlating the vast amounts of data coming forth from the studies of genomes without the foundations of Darwin's theory. As Theodosius Dobzhansky, a leading biologist of the twentieth century (and a devout Eastern Orthodox Christian), has said, "Nothing in biology makes sense except in the light of evolution."[3]

Clearly, however, evolution has been the source of great discomfort in the religious community over the past 150 years, and that resistance shows no signs of lessening. Yet believers would be well advised to look carefully at the overwhelming weight of scientific data supporting this view of the relatedness of all living things, including ourselves. Given the strength of the evidence, it is perplexing that so little progress in public acceptance has occurred in the United States. Perhaps part of the problem relates to a simple misunderstanding of the word "theory." Critics are fond of pointing out that evolution is "only a theory," a statement that puzzles working scientists who are used to a different meaning of that word. My Funk & Wagnalls dictionary provides the following two alternative definitions for the word "theory": "(1) a speculative or conjectural view of something; (2) fundamental principles underlying a science, art, etc.: music theory, theory of equations."

It is this second usage that scientists intend when they talk about evolutionary theory, just as when they mention gravitational theory or the germ theory of infectious disease. In this context, the word "theory" is not intended to convey uncertainty; for that purpose a scientist would use the word "hypothesis." In common, everyday usage, however, "theory" takes on a much more casual context, as reflected in Funk & Wagnalls option 1: "I have a theory that Bill has a crush on Mary," or "Linda's theory is that the butler did it." It is too bad that our language lacks the necessary subtleties of distinction here, as clearly this simple confusion of the meaning of the word has made things worse in the contentious dialogue between science and faith about how living things are related.

So if evolution is true, is there any place left for God? Arthur Peacocke, the distinguished British molecular biologist who subsequently became an Anglican priest and has written extensively about the interface between biology and faith, has recently published a book entitled *Evolution: The Disguised Friend of Faith?* That interesting title suggests a possible rapprochement, but is this a shotgun marriage of incompatible worldviews? Or, now that we have laid out the arguments for the plausibility of God, on the one hand, and the scientific data about the origins of the universe and life on our planet, on the other, can we find a happy and harmonious synthesis?

PART THREE

Faith in Science, Faith in God

CHAPTER SIX

Genesis, Galileo, and Darwin

W ASHINGTON, D.C., is full of smart, hard-driving, interesting people. A wide variety of religious faiths
are represented, as well as a significant proportion
of atheists and agnostics. When I was invited to speak at the
annual men's dinner at a highly regarded Protestant church
just outside the District, I gladly accepted. It was an inspiring
evening as prominent leaders, teachers, and blue-collar workers collectively let their hair down to talk earnestly about their
faith, and to ask penetrating questions about how science and
faith can contradict or reinforce each other. For a good hour of
discourse, goodwill filled the room. And then one church
member asked the senior pastor whether he believed that the
first chapter of Genesis was a literal, step-by-step, day-by-day
description of the origins of the earth and of humankind. In an

instant, brows furrowed and jaws tightened. Harmony re-
treated to the far corners of the room. The pastor's carefully
worded response, worthy of the most deft politician, managed
utterly to avoid answering the question. Most of the men
looked relieved that a confrontation had been avoided, but the
spell was broken.

A few months later I spoke to a national gathering of Christian physicians, explaining how I had found great joy in being
both a scientist studying the genome and a follower of Christ.
Warm smiles abounded; there was even an occasional "Amen."
But then I mentioned how overwhelming the scientific evidence
for evolution is, and suggested that in my view evolution might
have been God's elegant plan for creating humankind. The
warmth left the room. So did some of the attendees, literally
walking out, shaking their heads in dismay.

What's going on here? From a biologist's perspective, the
evidence in favor of evolution is utterly compelling. Darwin's
theory of natural selection provides a fundamental framework
for understanding the relationships of all living things. The predictions of evolution have been borne out in more ways than
Darwin could have possibly imagined when he proposed his
theory 150 years ago, especially in the field of genomics.

If evolution is so overwhelmingly supported by scientific evidence, then what are we to make of the lack of public support
for its conclusions? In 2004, the distinguished Gallup organization posed the following question to a statistical sample of
Americans: "Do you think that (1) Charles Darwin's theory of
evolution is a scientific theory that has been well supported by
evidence, or (2) just one of many theories and one that has not

been well supported by evidence, or (3) don't you know enough about it to say?" Only one-third of Americans indicated that they believed the theory of evolution was well supported, with the remainder being equally divided between those who argued it had not been well supported and those who just didn't know enough to say.

When the question was posed in a more explicit way to ask about the origin of human beings, an even stronger percentage seemed to reject the conclusions of evolution. Here's the question: "Which of the following statements comes closest to your views on the origin and development of human beings? (1) Human beings have developed over millions of years from less advanced forms of life, but God guided this process. (2) Human beings have developed over millions of years from less advanced forms of life, but God had no part in this process. (3) God created human beings pretty much in their present form at one time within the last 10,000 years or so."

In 2004, 45 percent of Americans chose option 3, 38 percent chose option 1, and 13 percent chose option 2. These statistics have remained essentially unchanged over the past twenty years.

REASONS FOR LACK OF PUBLIC ACCEPTANCE OF DARWIN'S THEORY

There can be no question that the theory of evolution is counterintuitive. For centuries, humans have been closely observing the natural world around them. Most observers, regardless of religious persuasion, have been unable to account for the complexity and diversity of life forms without postulating a designer.

Darwin's idea was revolutionary because it offered such a totally unexpected conclusion. Seeing new species evolve was not part of anyone's everyday experience. Despite the unquestioned complexity of certain inanimate objects (such as snowflakes), the complexity of life forms seemed wildly out of proportion to anything observable in the inanimate world. William Paley's parable of finding a watch on the moor—which would cause any of us to deduce the existence of a watchmaker—resonated with many readers in the ninteenth century, and continues to resonate with many people today. Life appears designed, so there must be a designer.

A major part of the problem in accepting the theory of evolution is that it requires one to grasp the significance of extremely long periods of time involved in the process. Such intervals are unimaginably beyond individual experience. One way to reduce the eons of history into a more comprehensible form is to imagine what would happen if the 4.5 billion years of the earth's existence, from initial formation to today, were instead compressed into a twenty-four-hour day. If the earth was formed at 12:01 A.M., then life would appear at about 3:30 A.M. After a long day of slow progression to multicellular organisms, the Cambrian explosion would finally occur at about 9 P.M. Later that evening, dinosaurs would roam the earth. Their extinction would occur at 11:40 P.M., at which time the mammals would begin to expand.

The divergence of branches leading to chimps and humans would occur with only one minute and seventeen seconds remaining in the day, and anatomically modern humans would appear with just three seconds left. The life of a middle-aged human on earth today would occupy only the last millisecond (one-thousandth of a second). It is not surprising that many of us have a great deal of difficulty contemplating evolutionary time.

Furthermore, there can be no question that the major resistance to broad public acceptance of evolution, especially in the United States, relates to the perception that it argues against the role of a supernatural designer. This objection, if true, has to be taken with great seriousness by all believers. If you are compelled (as I am) by the existence of the Moral Law and the universal longing for God, if you sense that there is a glowing signpost within our hearts, pointing toward a benevolent and loving presence, then it is quite natural to resist any force that seems bent upon chopping down the sign. Before mounting an all-out war against that invading force, however, we had best be certain that we are not shooting at a neutral observer, or maybe even an ally.

The problem for many believers, of course, is that the conclusions of evolution appear to contradict certain sacred texts that describe God's role in the creation of the universe, the earth, all living things, and ourselves. In Islam, for instance, the Qur'an describes life developing in stages, but sees humans as a special act of creation "from sounding clay, from mud molded into shape" (15:26). In Judaism and Christianity, the great creation story of Genesis 1 and 2 is a solid bedrock for many believers.

WHAT DOES GENESIS REALLY SAY?

If you have not recently read this Biblical account, find a Bible right now and read Genesis 1:1 through Genesis 2:7. There is no substitute for looking at the actual text if one is trying to understand its meaning. And if you are concerned that the words in this text have been seriously compromised by centuries of copying and recopying, do not worry very much about this—the evidence in favor of the authenticity of the Hebrew is in fact quite strong.

There is no question that this is a powerful and poetic narrative recounting the story of God's creative actions. "In the beginning God created the heavens and the earth" implies that God always existed. This description is certainly compatible with scientific knowledge of the Big Bang. The remainder of Genesis 1 describes a series of creative acts, from "Let there be light" on day one, to the waters and the sky on day two, to the appearance of land and vegetation on day three, the sun, moon, and stars on day four, fish and birds on day five, and finally on a very busy sixth day, the appearance of land animals and male and female humans.

Genesis 2 then begins with a description of God resting on the seventh day. After this appears a second description of the creation of humans, this time explicitly referring to Adam. The second creation description is not entirely compatible with the first; in Genesis 1 vegetation appears three days before humans are created, whereas in Genesis 2 it seems that God creates Adam from the dust of the earth before any shrub or plant had yet appeared. In Genesis 2:7, it is interesting to note, the

Hebrew phrase that we translate "living being" is applied to Adam in exactly the same way it was previously applied to fish, birds, and land animals in Genesis 1:20 and 1:24.

What are we to make of these descriptions? Did the writer intend for this to be a literal depiction of precise chronological steps, including days of twenty-four-hour duration (though the sun was not created until day three, leaving open the question of how long a day would have been before that)? If a literal description was intended, why then are there two stories that do not entirely mesh with each other? Is this a poetic and even allegorical description, or a literal history?

These questions have been debated for centuries. Nonliteral interpretations since Darwin are somewhat suspect in some circles, since they could be accused of "caving in" to evolutionary theory, and perhaps thereby compromising the truth of the sacred text. So it is useful to discover how learned theologians interpreted Genesis 1 and 2 long before Darwin appeared on the scene, or even before geologic evidence of the extreme age of the earth began to accumulate.

In that regard, the writings of Saint Augustine, a converted skeptic and brilliant theologian who lived around 400 A.D., are of particular interest. Augustine was fascinated by the first two chapters of Genesis, and wrote no less than five extensive analyses of these texts. Put down more than sixteen hundred years ago, his thoughts are still illuminating today. Reading through those intensely analytical musings, especially as recorded in *The Literal Meaning of Genesis,* the *Confessions,* and *The City of God,* it is clear that Augustine is posing more questions than he is providing answers for. He repeatedly re-

151

turns to the question of the meaning of time, concluding that God is outside of time and not bounded by it (2 Peter 3:8 states this explicitly: "With the Lord a day is like a thousand years, and a thousand years are like a day"). This in turn causes Augustine to question the duration of the seven days of biblical creation.

The Hebrew word used in Genesis 1 for day *(yôm)* can be used both to describe a twenty-four-hour day and to describe a more symbolic representation. There are multiple places in the Bible where *yôm* is utilized in a nonliteral context, such as "the day of the Lord"—just as we might say "in my grandfather's day" without implying that Grandpa had lived only twenty-four hours.

Ultimately, Augustine writes: "What kind of days these were, it is extremely difficult, or perhaps impossible for us to conceive."[1] He admits there are probably many valid interpretations of the book of Genesis: "With these facts in mind, I have worked out and presented the statements of the book of Genesis in a variety of ways according to my ability; and, in interpreting words that have been written obscurely for the purpose of stimulating our thought, I have not brashly taken my stand on one side against a rival interpretation which might possibly be better."[2]

Diverse interpretations continue to be promoted about the meaning of Genesis 1 and 2. Some, particularly from the evangelical Christian church, insist upon a completely literal interpretation, including twenty-four-hour days. Coupled with subsequent genealogical information in the Old Testament, this leads to Bishop Ussher's famous conclusion that God created heaven and earth in 4004 B.C. Other equally sincere believers do not accept the requirement that the days of creation need be twenty-four hours in length, though they otherwise accept the narrative as a literal and sequential depiction of God's creative acts. Still other believers see the language of Genesis 1 and 2 as intended to instruct readers of Moses' time about God's character, and not to attempt to teach scientific facts about the specifics of creation that would have been utterly confusing at the time.

Despite twenty-five centuries of debate, it is fair to say that no human knows what the meaning of Genesis 1 and 2 was precisely intended to be. We should continue to explore that! But the idea that scientific revelations would represent an enemy in that pursuit is ill conceived. If God created the universe, and the laws that govern it, and if He endowed human beings with intellectual abilities to discern its workings, would He want us to disregard those abilities? Would He be diminished or threatened by what we are discovering about His creation?

LESSONS FROM GALILEO

Watching the current fireworks between certain branches of the church and certain outspoken scientists, an observer with a sense of history might ask, "Haven't we been to this movie before?" Conflicts between interpretation of scripture and scientific observations are not exactly new. In particular, the conflicts that arose in the seventeenth century between the Christian church and the science of astronomy provide some instructive context for the evolutionary debates of today.

Galileo Galilei was a brilliant scientist and mathematician, born in Italy in 1564. Not satisfied to carry out mathematical analyses of other people's data, or to follow the Aristotelian tradition of posing theories without requiring experimental support, Galileo was involved in both experimental measurements and using mathematics to interpret them. In 1608, inspired by information he had heard about the invention of the telescope in the Netherlands, Galileo made his own instrument and quickly made a number of astronomical observations of profound significance. He observed four moons orbiting the planet Jupiter. That simple observation, which we take for granted today, presented significant problems for the traditional Ptolemaic system, where all heavenly bodies were supposed to rotate around the earth. Galileo also observed sunspots, which represented a possible affront to the idea that all heavenly bodies were created perfect.

Galileo ultimately came to the conclusion that his observations could make sense only if the earth revolved around the sun. That placed him in direct conflict with the Catholic Church.

While much of the traditional lore about Galileo's persecutions by the church is overblown, there is no question that his conclusions were received with alarm in many theological quarters. This was not entirely based on religious arguments, however. In fact, his observations were accepted by many Jesuit astronomers, but resented by rival academics, who urged the church to intervene. The Dominican Father Caccini obliged. In a sermon directly targeting Galileo, the friar insisted that "geometry is of the devil" and that "mathematicians should be banished as the authors of all heresies."[3]

Another Catholic priest claimed that Galileo's conclusions were not only heretical but atheistic. Other attacks included a claim that "his pretended discovery vitiates the whole Christian plan of salvation" and that "it casts suspicion on the doctrine of the incarnation."

In retrospect, modern observers must wonder why the church was so utterly threatened by the idea of the earth revolving around the sun. To be sure, certain verses from scripture seemed to support the church's position, such as Psalm 93:1—"The world is firmly established; it cannot be moved"—and Psalm 104:5: "He set the earth on its foundation; it can never be moved." Also cited was Ecclesiastes 1:5: "The sun rises and the sun sets, and hurries back to where it rises." Today, few believers argue that the authors of these verses were intending to teach science. Nonetheless, passionate claims were made to that effect, implying that a heliocentric system would somehow undermine the Christian faith.

Despite having upset the religious establishment, Galileo got by with a warning not to teach or defend his views. Subsequently, a new pope, who was friendly to Galileo, gave him vague permission to write a book about his opinions, so long as it provided a balanced view. Galileo's masterwork, *Dialogue Concerning the Two Chief World Systems*, presented a fanciful dialogue between a geocentric and a heliocentric enthusiast, moderated by a neutral but interested layman. The narrative frame fooled nobody. Galileo's preference for the heliocentric point of view was obvious by the end of the book, and despite its approval by Catholic censors, it caused an uproar.

Galileo was subsequently tried before the Roman Inquisition in 1633, and ultimately forced to "abjure, curse, and detest" his own work. He remained under house arrest for the remainder of his life, and his publications were banned. Only in 1992— 359 years after the trial—was an apology issued by Pope John Paul II: "Galileo sensed in his scientific research the presence of the Creator who, stirring in the depths of his spirit, stimulated him, anticipating and assisting his intuitions."[4]

So in this example, the scientific correctness of the heliocentric view ultimately won out, despite strong theological objections. Today all faiths except perhaps a few primitive ones seem completely at home with this conclusion. The claims that heliocentricity contradicted the Bible are now seen to have been overstated, and the insistence on a literal interpretation of those particular scripture verses seems wholly unwarranted.

Could this same harmonious outcome be realized for the current conflict between faith and the theory of evolution? On the positive side, the Galileo affair demonstrates that a contentious chapter did eventually get resolved on the basis of overwhelming scientific evidence. But along the way, considerable damage was done—and more to faith than to science. In his commentary on Genesis, Augustine provides an exhortation that might well have been heeded by the seventeenth-century church:

> Usually, even a non-Christian knows something
> about the earth, the heavens, and the other ele-

ments of this world, about the motion and orbit of the stars and even their size and relative positions, about the predictable eclipses of the sun and moon, the cycles of the years and the seasons, about the kinds of animals, shrubs, stones, and so forth, and this knowledge he holds to as being certain from reason and experience.

Now, it is a disgraceful and dangerous thing for an infidel to hear a Christian, presumably giving the meaning of Holy Scripture, talking nonsense on these topics; and we should take all means to prevent such an embarrassing situation, in which people show a vast ignorance in a Christian and laugh it to scorn.

The shame is not so much that an ignorant individual is derided, but the people outside the household of the faith think our sacred writers held such opinions, and, to the great loss of those for whose salvation we toil, the writers of our Scripture are criticized and rejected as unlearned men. If they find a Christian mistaken in a field which they themselves know well and hear him maintaining his foolish opinions about our books, how are they going to believe those books and matters concerning the resurrection of the dead, the hope of eternal life, and the kingdom of heaven, when they think their pages are full of falsehoods on facts which they themselves have learned from experience in the light of reason?[5]

Unfortunately, however, in many ways the controversy between evolution and faith is proving to be much more difficult than an argument about whether the earth goes around the sun. After all, the evolution controversy reaches into the very heart of both faith and science. This is not about rocky heavenly bodies, but about ourselves and our relation to a Creator. Perhaps the centrality of those issues explains the fact that, despite the modern rate of progress and dissemination of information, we still have not resolved the public controversy about evolution, nearly 150 years after Darwin's publication of *The Origin of Species.*

Galileo remained a strong believer to the end. He continued to argue that scientific exploration was not only an acceptable but a noble course of action for a believer. In a famous remark that could be the motto today of all scientist-believers, he said: "I do not feel obliged to believe that the same God who has endowed us with sense, reason, and intellect has intended us to forgo their use."[6]

Keeping that exhortation in mind, let us now explore the possible responses to the contentious interaction between the theory of evolution and faith in God. Each of us must come to some conclusion here, and choose one of the following positions. When it comes to the meaning of life, fence sitting is an inappropriate posture for both scientists and believers.

CHAPTER SEVEN

Option 1: Atheism and Agnosticism

(When Science Trumps Faith)

MY JUNIOR YEAR IN COLLEGE, 1968, was full of deeply troubling events. Soviet tanks had rolled into Czechoslovakia; the Vietnam War had escalated with the Tet offensive; and Robert F. Kennedy and Martin Luther King had been assassinated. But at the very end of that year, another much more positive event occurred that electrified the world—the launch of *Apollo 8*. It was the first manned spacecraft to orbit the moon. Frank Borman, James Lovell, and William Anders traveled through space for three days that December, while the world held its breath. Then they began to circle the moon, taking the first human photos of Earth rising over the moon's surface, reminding us all just how small and fragile our planet appears from the vantage point of space. On Christmas Eve, the three astronauts broadcast a live television transmission from their capsule. After commenting on their experiences and on the starkness of the lunar landscape, they jointly read to the world the first ten verses of Genesis 1. As an agnostic on the way to becoming an atheist at the time, I still remember the surprising sense of awe that settled over me as those unforgettable words—"In the beginning, God created the heavens and the earth"—reached my ears from 240,000 miles away, spoken by men who were scientists and engineers, but for whom these words had obvious powerful meaning.

Shortly afterward, the famous American atheist Madalyn Murray O'Hair filed suit against NASA for permitting this Christmas Eve reading of the Bible. She argued that U.S. astronauts, who are federal employees, should be banned from public prayer in space. Though the courts ultimately rejected her suit, NASA discouraged such references to faith in future flights. Thus, Buzz Aldrin of *Apollo 11* arranged to take communion on the surface of the moon during the first human lunar landing in 1969, but that event was never publicly reported.

A militant atheist taking legal action against a Bible reading by astronauts circling the moon on Christmas Eve: what a symbol of the escalating hostility between believers and nonbelievers in our modern world! No one objected in 1844 when Samuel Morse's first telegraph message was "What hath God wrought?" Yet increasingly in the twenty-first century, extremists on both sides of the science/faith divide are insisting that the other be silenced.

Atheism has evolved in the decades since O'Hair was its most visible advocate. Today, it is not secular activists like O'Hair who make up its vanguard—it is evolutionists. Among several vocal proponents, Richard Dawkins and Daniel Dennett stand out as articulate academics who expend considerable energies to explain and extend Darwinism, proclaiming that an acceptance of evolution in biology requires an acceptance of atheism in theology. In a remarkable marketing ploy, they and their colleagues in the atheist community have also attempted to promote the term "bright" as an alternative to "atheist." (The implied corollary, that believers must be "dim," may be one good reason why the term has yet to catch on.) Certainly, their hostility to belief is undisguised. How did we get here?

ATHEISM

Some have divided atheism into "weak" and "strong" forms. Weak atheism is the absence of belief in the existence of a God or gods, whereas strong atheism is the firm conviction that no such deities exist. In everyday conversation, strong atheism is generally the assumed position of someone who takes this point of view, and so I will consider that perspective here.

Elsewhere I have argued that the search for God is a broadly shared attribute of all humankind, across geographic areas and throughout human history. In his remarkable book *Confessions* (essentially the first Western autobiography), Saint Augustine describes this longing in the very first paragraph: "Nevertheless, to praise you is the desire of man, a little piece of your creation. You stir man to take pleasure in praising you, because you have made us for yourself, and our heart is restless until it rests in you."[1]

If this universal search for God is so compelling, what are we to make of those restless hearts who deny His existence? On what foundation do they make such assertions with such confidence? And what are the historic origins of this point of view?

Atheism played a minor role in human history until the eighteenth century, with the advent of the Enlightenment and the rise of materialism. But it was not just the discovery of natural laws that opened the door to an atheistic perspective; after all, Sir Isaac Newton was a firm believer in God, and wrote and published more works on interpretation of the Bible than on mathematics and physics. A more powerful force giving rise to atheism in the eighteenth century was a rebellion against the oppressive authority of the government and the church, particularly as manifested in the French Revolution. Both the French royal family and the church leadership were seen as harsh, self-promoting, hypocritical, and insensitive to the needs of the common man. Equating the organized church with God Himself, revolutionaries decided it was better to throw off both.

Additional fuel for the atheist perspective subsequently was supplied by the writings of Sigmund Freud, who argued that belief in God is just wishful thinking. But even stronger support for the atheist perspective in the last 150 years has been seen to arise from Darwin's theory of evolution. Dismantling the "argument from design" that had been such a powerful arrow in the theist's quiver, the advent of evolutionary theory was seized upon by atheists as a powerful counterweapon against spirituality.

Consider, for instance, Edward O. Wilson, one of the most

outstanding evolutionary biologists of our time. In his book *On Human Nature,* Wilson cheerfully announces that evolution has triumphed over supernaturalism of any sort, concluding: "The final decisive edge enjoyed by scientific naturalism will come from its capacity to explain traditional religion, its chief competition, as a wholly material phenomenon. Theology is not likely to survive as an independent intellectual discipline."[2] Strong words.

Even stronger words have emanated from Richard Dawkins. In a series of books beginning with *The Selfish Gene* and extending through *The Blind Watchmaker, Climbing Mount Improbable,* and *A Devil's Chaplain,* Dawkins outlines with compelling analogies and rhetorical flourishes the consequences of variation and natural selection. Standing on this Darwinian foundation, Dawkins then extends his conclusions to religion in highly aggressive terms: "It is fashionable to wax apocalyptic about the threat to humanity posed by the AIDS virus, 'mad cow' disease, and many others, but I think a case can be made that faith is one of the world's great evils, comparable to the smallpox virus but harder to eradicate."[3]

In his recent book *Dawkins' God,* molecular biologist and theologian Alister McGrath takes on these religious conclusions and points out the logical fallacies behind them. Dawkins's arguments come in three main flavors. First, he argues that evolution fully accounts for biological complexity and the origins of humankind, so there is no more need for God. While this argument rightly relieves God of the responsibility for multiple acts of special creation for each species on the planet, it certainly does not disprove the idea that God worked out His creative plan by means of evolution. Dawkins's first argument is thus irrelevant to the God that Saint Augustine worshiped, or that I worship. But Dawkins is a master of setting up a straw man, and then dismantling it with great relish. In fact, it is hard to escape the conclusion that such repeated mischaracterizations of faith betray a vitriolic personal agenda, rather than a reliance on the rational arguments that Dawkins so cherishes in the scientific realm.

The second objection from the Dawkins school of evolutionary atheism is another straw man: that religion is antirational. He seems to have adopted the definition of religion attributed to Mark Twain's apocryphal schoolboy, "Faith is believing what you know ain't so."[4] Dawkins's definition of faith is "blind trust, in the absence of evidence, even in the teeth of evidence."[5] That certainly does not describe the faith of most serious believers throughout history, nor of most of those in my personal acquaintance. While rational argument can never conclusively prove the existence of God, serious thinkers from Augustine to Aquinas to C. S. Lewis have demonstrated that a belief in God is intensely plausible. It is no less plausible today. The caricature of faith that Dawkins presents is easy for him to attack, but it is not the real thing.

Dawkins's third objection is that great harm has been done in the name of religion. There is no denying this truth, though undeniably great acts of compassion have also been fueled by faith. But evil acts committed in the name of religion in no way impugn the truth of the faith; they instead impugn the nature of human beings, those rusty containers into which the pure water of that truth has been placed.

Interestingly, while Dawkins argues that it is the gene and

its relentless drive for survival that explains the existence of all living things, he argues that we humans are at last far enough advanced to be able to rebel against our genetic imperatives. "We can even discuss ways of deliberately cultivating and nurturing pure, disinterested altruism—something that has no place in nature, something that has never existed before in the whole history of the world."[6]

Now here is a paradox: apparently Dawkins is a subscriber to the Moral Law. Where might this rush of good feeling have come from? Surely this should arouse Dawkins's suspicion about the "blind pitiless indifference" that he argues is conferred on all of nature, including himself and all the rest of humankind, by godless evolution? What value should he then attach to altruism?

The major and inescapable flaw of Dawkins's claim that science demands atheism is that it goes beyond the evidence. If God is outside of nature, then science can neither prove nor disprove His existence. Atheism itself must therefore be considered a form of blind faith, in that it adopts a belief system that cannot be defended on the basis of pure reason. Perhaps the most colorful encapsulation of this point of view comes from an unlikely source: Stephen Jay Gould, who outside of Dawkins is probably the most widely read public spokesperson for evolution of the past generation. Writing in an otherwise little-noticed book review, Gould chastised the Dawkins perspective:

> To say it for all my colleagues and for the umpteenth millionth time: Science simply cannot by its legitimate methods adjudicate the issue of God's

possible superintendence of nature. We neither affirm nor deny it; we simply can't comment on it as scientists. If some of our crowd have made untoward statements claiming that Darwinism disproves God, then I will find Mrs. McInerney [Gould's third-grade teacher] and have their knuckles rapped for it. . . . Science can work only with naturalistic explanations; it can neither affirm nor deny other types of actors (like God) in other spheres (the moral realm, for example). Forget philosophy for a moment; the simple empirics of the past hundred years should suffice. Darwin himself was agnostic (having lost his religious beliefs upon the tragic death of his favorite daughter), but the great American botanist Asa Gray, who favored natural selection and wrote a book entitled *Darwiniana,* was a devout Christian. Move forward 50 years: Charles D. Walcott, discoverer of the Burgess Shale Fossils, was a convinced Darwinian and an equally firm Christian, who believed that God had ordained natural selection to construct the history of life according to His plans and purposes. Move on another 50 years to the two greatest evolutionists of our generation: G. G. Simpson was a humanistic agnostic, Theodosius Dobzhansky, a believing Russian Orthodox. Either half my colleagues are enormously stupid, or else the science of Darwinism is fully compatible with conventional religious beliefs—and equally compatible with atheism.[7]

So those who choose to be atheists must find some other basis for taking that position. Evolution won't do.

AGNOSTICISM

The term "agnostic" was coined by the colorful British scientist Thomas Henry Huxley, also known as "Darwin's bulldog," in 1869. Here is his description of how he came to originate the term:

> When I reached intellectual maturity, and began to ask myself whether I was an atheist, a theist, or a pantheist; a materialist or an idealist; a Christian or a free thinker, I found that the more I learned and reflected, the less ready was the answer; until at last I came to the conclusion that I had neither art nor part with any of these denominations, except the last. The one thing in which most of these good people were agreed was the one thing in which I differed from them. They were quite sure that they had attained a certain "gnosis"—had more or less successfully solved the problem of existence; while I was quite sure I had not, and had a pretty strong conviction that the problem was insoluble. . . . So I took thought, and invented what I conceived to be the appropriate title of "agnostic." It came into my head as suggestively antithetic to the "gnostic" of church history, who professed to know so much about the very things of which I was ignorant.[8]

An agnostic, then, is one who would say that the knowledge of God's existence simply cannot be achieved. As with atheism, there are strong and weak forms of agnosticism, with the strong form indicating there is no way that humankind will ever know, whereas the weak form simply says, "Not now."

The boundary lines between strong agnosticism and weak atheism are blurry, as an interesting Darwin anecdote reveals. At a dinner party with two atheists in 1881, Darwin asked his guests, "Why do you call yourselves atheists?" saying that he preferred Huxley's word, "agnostic." One of his guests replied that "agnostic was but atheist writ respectable, and atheist was only agnostic writ aggressive."[9]

Most agnostics, however, are not so aggressive, and simply take the position that it is not possible, at least not for them at that time, to take a position for or against the existence of God. On the surface, this is a logically defensible position (whereas atheism is not). Certainly it is entirely compatible with the theory of evolution, and many biologists would put themselves in this camp. But agnosticism also runs the risk of being a cop-out.

To be well defended, agnosticism should be arrived at only after a full consideration of all of the evidence for and against the existence of God. It is a rare agnostic who has made the effort to do so. (Some who have, and a rather distinguished list it is, have unexpectedly converted themselves to belief in God.) Furthermore, while agnosticism is a comfortable default pattern for many, from an intellectual perspective it conveys a certain tinniness. Would we admire someone who insisted the age of the universe was unknowable, and hadn't taken time to look at the evidence?

CONCLUSION

Science cannot be used to justify discounting the great monotheistic religions of the world, which rest upon centuries of history, moral philosophy, and the powerful evidence provided by human altruism. It is the height of scientific hubris to claim otherwise. But that leaves us with a challenge: if the existence of God is true (not just tradition, but actually true), and if certain scientific conclusions about the natural world are also true (not just in fashion, but objectively true), then they cannot contradict each other. A fully harmonious synthesis must be possible.

Looking at our current world, however, it is hard to escape the sense that these two versions of truth are not seeking harmony, but are at war. That is nowhere more apparent than in the debates over Darwin's theory of evolution. Here is where the battles are currently most furiously raging; here is where the misunderstanding on both sides is most profound; here is where the stakes for our future world are the highest; and here is where harmony is most desperately needed. So here is where we had best turn our attention next.

CHAPTER EIGHT

Option 2: Creationism

(When Faith Trumps Science)

F EW RELIGIOUS OR SCIENTIFIC VIEWS can be neatly summed up in a single word. The application of misleading labels for particular perspectives has regularly muddied the debate between science and faith throughout the modern era. Nowhere is this more true than in the case of the "creationist" label, which has featured so prominently in the science-and-faith debates over the past century. Taken at face value, the term "creationist" would seem to imply the general perspective of one who argues for the existence of a God who was directly involved in the creation of the universe. In that broad sense, many deists and nearly all theists, including me, would need to count themselves as creationists.

YOUNG EARTH CREATIONISM

Over the past century, however, the term "Creationist" has been hijacked (and capitalized) to apply to a very specific subset of such believers, specifically those who insist on a literal reading of Genesis 1 and 2 to describe the creation of the universe and the formation of life on earth. The most extreme version of this view, generally referred to as Young Earth Creationism (YEC), interprets the six days of creation as literal twenty-four-hour days and concludes that the earth must be less than ten thousand years old. YEC advocates also believe that all species were created by individual acts of divine creation, and that Adam and Eve were historical figures created by God from dust in the Garden of Eden, and not descended from other creatures.

YEC believers generally accept the idea of "microevolution," whereby small changes within species can occur by variation and natural selection, but they reject the concept of "macroevolution," the process that would allow one species to evolve into another. They argue that perceived gaps in the fossil record demonstrate the fallacy of Darwin's theory. In the 1960s, the YEC movement was further crystallized by publication of *The Genesis Flood* and subsequent writings of members of the Institute for Creation Research, founded by the late Henry Morris. Among the many assertions of Morris and his colleagues were that the geologic strata and the fossils within the various layers were created in a few weeks by the worldwide flood described in Genesis 6–9, rather than having been deposited over hundreds of millions of years. Judging by polls, Young Earth Creationism is the view held by approximately 45 percent of Americans. Many evangelical

Christian churches are closely aligned with this view. Many books and videos can be found in Christian bookstores that claim that no intermediate fossil forms can be found for birds, turtles, elephants, or whales (yet examples of all of these have been found in the last few years), that the Second Law of Thermodynamics rules out the possibility of evolution (it clearly does not), and that radioactive dating of rocks and the universe is wrong because decay rates have changed over time (they have not). One can even visit Creationist museums and theme parks that depict humans frolicking with dinosaurs, since the YEC perspective does not accept the idea that dinosaurs became extinct long before humans appeared on the scene.

Young Earth Creationists argue that evolution is a lie. They postulate that the relatedness of organisms as visualized by the study of DNA is simply a consequence of God having used some of the same ideas in His multiple acts of special creation. Confronted with such facts as the similar ordering of genes across chromosomes between different mammalian species, or the existence of repetitive "junk DNA" in shared locations along the DNA of humans and mice, YEC advocates simply dismiss this as part of God's plan.

YOUNG EARTH CREATIONISM AND MODERN SCIENCE ARE INCOMPATIBLE

In general, those who hold these views are sincere, well-meaning, God-fearing people, driven by deep concerns that naturalism is threatening to drive God out of human experience. But the claims of Young Earth Creationism simply cannot be accommodated by tinkering around the edges of scientific knowledge. If these claims were actually true, it would lead to a complete and irreversible collapse of the sciences of physics, chemistry, cosmology, geology, and biology. As biology professor Darrel Falk points out in his wonderful book *Coming to Peace with Science*, written specifically from his perspective as an evangelical Christian, the YEC perspective is the equivalent of insisting that two plus two is really not equal to four.

For anyone familiar with the scientific evidence, it is almost incomprehensible that the YEC view has achieved such wide support, especially in a country like the United States that claims to be so intellectually advanced and technologically sophisticated. But YEC advocates are serious about their faith first and foremost, and deeply concerned about a trend toward non-literal interpretations of the Bible, which might ultimately dilute the power of the scriptures to teach reverence for God to humankind. Young Earth Creationists argue that accepting anything other than acts of special divine creation during the six twenty-four-hour days of Genesis 1 would put the believer on a slippery slope toward a counterfeit faith. This argument appeals to the strong and understandable instincts of serious believers that their first priority is allegiance to God, and that apparent attacks on His person are to be fended off vigorously.

But Ultraliteral Interpretations of Genesis Are Unnecessary

Harkening back to Saint Augustine's interpretation of Genesis 1 and 2, however, and remembering that he had no reason to be accommodating to scientific evidence about evolution or the age of the earth, it is clear that the ultraliteral YEC views are in fact not required by a careful, sincere, and worshipful reading of the original text. In fact, this narrow interpretation is largely a creation of the last hundred years, arising in large consequence as a reaction to Darwinian evolution.

The concern about not accepting liberal interpretations of biblical texts is understandable. After all, there are clearly parts of the Bible that are written as eyewitness accounts of historical events, including much of the New Testament. For a believer, the events recorded in these sections ought to be taken as the writer intended—as descriptions of observed facts. But other parts of the Bible, such as the first few chapters of Genesis, the book of Job, the Song of Solomon, and the Psalms, have a more lyrical and allegorical flavor, and do not generally seem to carry the marks of pure historical narrative. To Saint Augustine, and to most other interpreters throughout history, until Darwin put believers on the defensive, the first chapters of Genesis had much more the feel of a morality play than an eyewitness report on the evening news.

The insistence that every word of the Bible must be taken literally runs into other difficulties. Surely the right arm of God did not really lift up the nation of Israel (Isaiah 41:10). Surely it is not part of God's nature to become forgetful and to need to be reminded of important matters from time to time by the prophets (Exodus 33:13). The intention of the Bible was (and is) to reveal the nature of God to humankind. Would it have served God's purposes thirty-four hundred years ago to lecture to His people about radioactive decay, geologic strata, and DNA?

Many believers in God have been drawn to Young Earth Creationism because they see scientific advances as threatening to God. But does He really need defending here? Is not God the author of the laws of the universe? Is He not the greatest scientist? The greatest physicist? The greatest biologist? Most important, is He honored or dishonored by those who would demand that His people ignore rigorous scientific conclusions about His creation? Can faith in a loving God be built on a foundation of lies about nature?

GOD AS THE GREAT DECEIVER?

Assisted by Henry Morris and colleagues, Young Earth Creationism has in the last half century attempted to provide alternative explanations for the wealth of observations about the natural world that seem to contradict the YEC position. But the fundamentals of so-called scientific Creationism are hopelessly flawed. Recognizing the overwhelming body of scientific evidence, some YEC advocates have more recently taken the tack of arguing that all of this evidence has been designed by God to mislead us, and therefore to test our faith. According to this argument, all of the radioactive decay clocks, all the fossils, and all of the genome sequences have been intentionally designed so it would look as if the universe was old, even though it was really created less than ten thousand years ago.

As Kenneth Miller points out in his excellent book, *Finding Darwin's God,* for these claims to be true, God would have had to engage in massive subterfuge. For instance, since many of the observable stars and galaxies in the universe are more than ten thousand light-years away, a YEC perspective would demand that our ability to observe them could come about only if God had fashioned all of those photons to arrive here in a "just so" fashion, even though they represent wholly fictitious objects.

This image of God as a cosmic trickster seems to be the ultimate admission of defeat for the Creationist perspective. Would God as the great deceiver be an entity one would want to worship? Is this consistent with everything else we know about God from the Bible, from the Moral Law, and from every other source—namely, that He is loving, logical, and consistent?

Thus, by any reasonable standard, Young Earth Creationism has reached a point of intellectual bankruptcy, both in its science and in its theology. Its persistence is thus one of the great puzzles and great tragedies of our time. By attacking the fundamentals of virtually every branch of science, it widens the chasm between the scientific and spiritual worldviews, just at a time where a pathway toward harmony is desperately needed. By sending a message to young people that science is dangerous, and that pursuing science may well mean rejecting religious faith, Young Earth Creationism may be depriving science of some of its most promising future talents.

But it is not science that suffers most here. Young Earth Creationism does even more damage to faith, by demanding that belief in God requires assent to fundamentally flawed claims about the natural world. Young people brought up in homes and churches that insist on Creationism sooner or later encounter the overwhelming scientific evidence in favor of an ancient universe and the relatedness of all living things through the process of evolution and natural selection. What a terrible and unnecessary choice they then face! To adhere to the faith of their childhood, they are required to reject a broad and rigorous body of scientific data, effectively committing intellectual suicide. Presented with no other alternative than Creationism, is it any wonder that many of these young people turn away from faith, concluding that they simply cannot believe in a God who would ask them to reject what science has so compellingly taught us about the natural world?

A PLEA FOR REASON

Let me conclude this brief chapter, therefore, with a loving entreaty to the evangelical Christian church, a body that I consider myself a part of, and that has done so much good in so many other ways to spread the good news of God's love and grace. As believers, you are right to hold fast to the concept of God as Creator; you are right to hold fast to the truths of the Bible; you are right to hold fast to the conclusion that science offers no answers to the most pressing questions of human existence; and you are right to hold fast to the certainty that the claims of atheistic materialism must be steadfastly resisted. But those battles cannot be won by attaching your position to a flawed foundation. To continue to do so offers the opportunity for the opponents of faith (and there are many) to win a long series of easy victories.

Benjamin Warfield, a conservative Protestant theologian in

the late nineteenth and early twentieth century, was well aware of the need for believers to stand firm in the eternal truths of their faith, despite great social and scientific upheavals. Yet he saw also the need to celebrate discoveries about the natural world that God created. Warfield wrote these remarkable words, which could well be embraced by the church today:

> We must not, then, as Christians, assume an attitude of antagonism toward the truths of reason, or the truths of philosophy, or the truths of science, or the truths of history, or the truths of criticism. As children of the light, we must be careful to keep ourselves open to every ray of light. Let us, then, cultivate an attitude of courage as over against the investigations of the day. None should be more zealous in them than we. None should be more quick to discern truth in every field, more hospitable to receive it, more loyal to follow it, whithersoever it leads.[1]

CHAPTER NINE

Option 3: Intelligent Design
(When Science Needs Divine Help)

THE YEAR 2005 WAS A TUMULTUOUS ONE for Intelligent Design theory, or ID as it is commonly known. The president of the United States gave it a partial endorsement, by stating that he thought schools should include this point of view when discussing evolution. His comment was made as a lawsuit against the school board of Dover, Pennsylvania, over a similar policy was heading to a much-ballyhooed trial. The media responded. Featured in cover stories of *Time* and *Newsweek*, discussed extensively on public radio and even on the front page of the *New York Times*, the controversy and confusion about ID escalated week by week. I found myself talking about it with scientists and editors, and even congressional staffers. In the fall, before the Dover trial was decided in favor of the plaintiffs, the citizens of Dover voted all of the members of their school board who had supported ID out of office.

Not since the 1925 Scopes trial has attention turned so intensively in the United States to a debate about evolution and its implications for religious faith. Perhaps this should be seen as a good thing—better to have an open debate than an underground attack on one point of view or another. But to most serious scientists who are committed believers, and even to some strong proponents of ID, things were getting seriously out of hand.

What Is Intelligent Design Anyway?

In its brief fifteen-year history, the ID movement has emerged as a major flash point for public discourse. Yet there remains much confusion about the basic tenets of this new entry on the scene.

First of all, just as with the term "creationism," there is a significant semantic difficulty. The two words "intelligent design" appear to encompass a broad range of interpretations of how life came to arise on this planet, and the role that God might have played in that process. But "Intelligent Design" (with capital letters) has become a term of art carrying a very specific set of conclusions about nature, especially the concept of "irreducible complexity." An observer unaware of this history might expect that anyone who believes in a God who cares about human beings (that is, a theist) would be someone who believes in Intelligent Design. But in the sense of current terminology, that would in most instances not be correct.

Intelligent Design burst on the scene in 1991. Some of its roots can be traced to earlier scientific arguments pointing out the statistical improbability of the origins of life. But ID places its major focus not on how the first self-replicating organisms came to be, but rather on perceived failings of the evolutionary theory to account for life's subsequent stunning complexity.

ID's founder is Phillip Johnson, a Christian lawyer at the University of California at Berkeley, whose book *Darwin on Trial* first laid out the ID position. Those arguments have been further expanded by others, especially Michael Behe, a biology professor whose book *Darwin's Black Box* elaborated the concept of irreducible complexity. More recently, William Dembski, a mathematician trained in information theory, has taken up a leading role as expositor of the ID movement.

The emergence of ID coincided with a series of judicial defeats to the teaching of creationism in U.S. schools, a chronological context that has caused critics to refer to ID uncharitably as "stealth creationism" or "creationism 2.0." But these terms do not do justice to the thoughtfulness and sincerity of ID's proponents. From my perspective as a geneticist, a biologist, and a believer in God, this movement deserves serious consideration.

The Intelligent Design movement basically rests upon three propositions.

Proposition 1: Evolution promotes an atheistic worldview and therefore must be resisted by believers in God.

Phillip Johnson, the founder, was driven not so much by a scientific desire to understand life (he makes no claim to be a scientist), but by a personal mission to defend God against what he perceived as growing public acceptance of a purely materialistic worldview. This concern finds much resonance in the faith community, where the triumphalist pronouncements of some of the most outspoken evolutionists of the day have led to a sense that some scientifically respectable alternative must be identified at all costs. (In that regard, ID could be thought of ironically as the rebellious love child of Richard Dawkins and Daniel Dennett.)

Johnson is quite forthright about his intentions, as laid out in his book *The Wedge of Truth: Splitting the Foundations of Naturalism*. The Discovery Institute, a major supporter of the ID movement, and for which Johnson serves as program adviser, carried this one step further in their "wedge document," which was originally intended as an internal memorandum but found its way onto the Internet. The document outlines five-, ten-, and twenty-year goals to influence public opinion, to effect an overthrow of atheistic materialism, and to replace it with a "broadly theistic understanding of nature."

Thus, while ID is presented as a scientific theory, it is fair to say that it was not born from the scientific tradition.

Proposition 2: Evolution is fundamentally flawed, since it cannot account for the intricate complexity of nature.

Students of history will recall that the argument that complexity requires a designer is the same one presented by William Paley in the early nineteenth century, and that Darwin himself found this logic quite compelling before arriving at his own explanation of evolution by natural selection. For the ID movement, however, this perspective has been dressed up in new clothes, namely the sciences of biochemistry and cell biology.

In *Darwin's Black Box,* Michael Behe outlines these arguments quite persuasively. When Behe the biochemist peers into the inner workings of the cell, he is amazed and awed (as am I) by the intricacies of the molecular machines that reside there, which science has been uncovering over the last several decades. There are elegant machines that translate RNA into protein, others that help the cell move around, and others that transmit signals from the cell surface to the nucleus, traveling along a cascading pathway of multiple components.

It is not just the cell that provides amazement. Entire organs, made up of billions or trillions of cells, are constructed in a way that can only inspire awe. Consider, for instance, the human eye, a complex cameralike organ whose anatomy and physiology continue to impress even the most sophisticated student of optics.

Behe argues that machines of this sort could never have arisen on the basis of natural selection. His arguments are focused primarily on complex structures that involve the interaction of multiple proteins, and whose function is lost if any one of those proteins is inactivated.

A particularly prominent example cited by Behe is the bacterial flagellum. Many different bacteria possess these flagella, which are little "outboard motors" that propel cells in various directions. The structure of the flagellum, which consists of about thirty different proteins, is really quite elegant. It includes miniature versions of a base anchor, a drive shaft, and a universal joint. All of this drives a filament propeller. The whole arrangement is a nanotechnology engineering marvel.

If any one of these thirty proteins is inactivated by genetic mutation, the whole apparatus will fail to work properly. Behe's argument is that such a complex device could never have come into being on the basis of Darwinian processes alone. He postulates that one component of this complex outboard motor might have evolved by chance over a long period of time, but there would have been no selective pressure to maintain it unless the other twenty-nine developed at the same time. Yet none of those would have enjoyed any selective advantage either until the entire structure had been assembled. Behe argues, and Dembski has later converted this to a more mathematical argument, that the probability of such accidental coevolution of multiple independently useless components is almost infinitely small.

Thus, the main scientific argument of the ID movement constitutes a new version of Paley's "argument from personal incredulity," now expressed in the language of biochemistry, genetics, and mathematics.

Proposition 3: If evolution cannot explain irreducible complexity, then there must have been an intelligent designer involved somehow, who stepped in to provide the necessary components during the course of evolution.

The ID movement is careful not to specify who this designer might have been, but the Christian perspective of most of the leaders of this movement implicitly suggests that this missing force would come from God Himself.

SCIENTIFIC OBJECTIONS TO ID

On the surface, the objections to Darwinism put forward by the ID movement appear compelling, and it is not surprising that nonscientists, especially those looking for a role for God in the evolutionary process, have embraced these arguments warmly. But if the logic truly had merit on scientific grounds, one would expect that the rank and file of working biologists would also show interest in pursuing these ideas, especially since a significant number of biologists are also believers. This has not happened, however, and Intelligent Design remains a fringe activity with little credibility within the mainstream scientific community.

Why is this so? Is this because, as ID proponents suggest, biologists are so used to worshiping at Darwin's altar that they cannot consider an alternative view? Since scientists are actually attracted to disruptive ideas, always looking for an opportunity to overturn accepted theories of the day, it seems unlikely that they would reject the arguments of ID simply because they challenge Darwin. In fact, the basis of the rejection is much more significant.

First of all, Intelligent Design fails in a fundamental way to qualify as a scientific theory. All scientific theories represent a framework for making sense of a body of experimental observations. But the primary utility of a theory is not just to look back but to look forward. A viable scientific theory predicts other findings and suggests approaches for further experimental verification. ID falls profoundly short in this regard. Despite its appeal to many believers, therefore, ID's proposal of the intervention of supernatural forces to account for complex multi-component biological entities is a scientific dead end. Outside of the development of a time machine, verification of the ID theory seems profoundly unlikely.

Core ID theory, as outlined by Johnson, also suffers by providing no mechanism by which the postulated supernatural interventions would give rise to complexity. In one attempt to address this, Behe has suggested that primitive organisms might have been "preloaded" with all of the genes that would ultimately be necessary for the development of the complex multi-component molecular machines that he considers irreducibly complex. Behe proposes that these sleeping genes were then awakened at an appropriate time hundreds of millions of years later, when they were needed. Setting aside the fact that no primitive organism can be found today that contains this cache of genetic information for future use, our knowledge of the mutational rate of genes that are not being utilized makes it highly improbable that such a storehouse of information would have survived long enough to be of any use.

Of even greater significance for the future of ID, it now seems likely that many examples of irreducible complexity are not irreducible after all, and that the primary scientific argument for ID is thus in the process of crumbling. In the short fifteen years since ID appeared on the scene, science has made substantial advances, particularly in the detailed study of the genomes of multiple organisms from multiple different parts of the evolutionary tree. Major cracks are beginning to appear, suggesting that ID proponents have made the mistake of confusing the unknown with the unknowable, or the unsolved with the unsolvable. Many books and articles have appeared on this topic,[1] and the interested reader is referred to those more explicit (and more technical) aspects of the debate. But here are three examples where structures that appeared to fit Behe's definition of irreducible complexity are clearly showing signs of

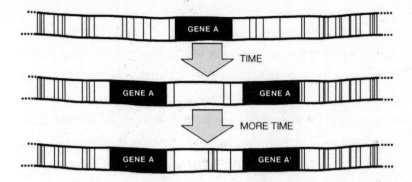

Figure 9.1 Evolution of a multiprotein complex by gene duplication. In the simplest circumstance, gene A provides an essential function to the organism. Duplication of that gene (an event that occurs frequently as genomes evolve) then creates a new copy. This copy is not essential for function (A is still providing that), so it is free to evolve without constraint. Rarely, a randomly arising small change will allow it to take on a new function (A') that is advantageous to the organism, resulting in positive selection. From a detailed study of DNA sequences, many complex multicomponent systems like the human clotting cascade appear to have arisen by this mechanism.

how they could have been assembled by evolution in a gradual step-by-step process.

The *human blood-clotting cascade,* appearing, with its dozen or more proteins, to be a complex system that Behe deems worthy of Rube Goldberg, can in fact be understood as the gradual recruitment of more and more elements of the cascade. The system appears to have begun with a very simple mechanism that would work satisfactorily for a low-pressure, low-flow hemodynamic system, and to have evolved over a long period of time into the complicated system necessary for humans and other mammals that have a high-pressure cardiovascular system, where leaks must be quickly stopped.

An important feature of this evolutionary hypothesis is the well-established phenomenon of gene duplication (Figure 9.1).

When one looks closely at the proteins in the clotting cascade, most of the components turn out to be related to one another at the level of amino acid sequence. This is not because wholly new proteins were constructed out of random genetic information and ultimately converged on the same theme. Rather, the similarity of these proteins can be shown to reflect ancient gene duplications that then allowed the new copy, unfettered by a need to maintain its original function (since the old copy was still doing that), to gradually evolve to take on a new function, driven by the force of natural selection.

Admittedly, we cannot precisely outline the order of the steps that ultimately led to the human clotting cascade. We may never be able to do so, because the host organisms of many predecessor cascades are lost to history. Yet Darwinism predicts that plausible intermediate steps must have existed, and some have indeed already been found. ID is silent on such predictions, and ID's central premise that the entire blood-clotting cascade had to emerge fully functional from prior DNA gibberish sets forth a straw man scenario that no serious student of biology would accept.[2]

The *eye* is another example frequently cited by advocates of Intelligent Design as displaying a degree of complexity that stepwise natural selection could never have achieved. Darwin himself recognized the difficulty that his readers would have accepting this: "To suppose that the eye with all of its inimitable contrivances for adjusting the focus to different distances, for admitting different amounts of light, and for the correction of spherical and chromatic aberration, could have been formed by natural selection, seems, I freely confess, absurd in the highest degree."[3] Yet Darwin, ever the impressive comparative biologist, proposed 150 years ago a series of steps in the evolution of this complex organ, which modern molecular biology is rapidly confirming.

Even very simple organisms have light sensitivity, which helps them avoid predators and seek food. Flatworms possess a simple pigmented pit, containing light-sensitive cells that provide some directionality to their ability to perceive incoming photons. The elegant chambered nautilus sports a modest advance, where this pit has been converted into a cavity with just a pinhole to admit light. This considerably improves the resolution of the apparatus, without requiring more than a subtle change in the geometry of the surrounding tissue. Similarly, the addition of a jellylike substance overlying the primitive light-sensitive cells in other organisms enables some focusing of the light. It is not prohibitively difficult, given hundreds of millions of years, to contemplate how this system could have evolved into the modern mammalian eye, complete with light-sensing retina and light-focusing lens.

It is also important to point out that the design of the eye does not appear on close inspection to be completely ideal. The rods and cones that sense light are the bottom layer of the retina, and light has to pass through the nerves and blood vessels to reach them. Similar imperfections of the human spine (not optimally designed for vertical support), wisdom teeth, and the curious persistence of the human appendix also seem to many anatomists to defy the existence of truly intelligent planning of the human form.

A particularly damaging crack in the foundation of Intelligent Design theory arises from recent revelations about the poster child of ID, the *bacterial flagellum*. The argument that it is irreducibly complex rests upon the presumption that the individual subunits of the flagellum could have had no prior useful function of some other sort, and therefore the motor could not have been assembled by recruiting such components in a stepwise fashion, driven by the forces of natural selection.

Recent research has fundamentally undercut this position.[4] Specifically, comparison of protein sequences from multiple bacteria has demonstrated that several components of the flagellum are related to an entirely different apparatus used by certain bacteria to inject toxins into other bacteria that they are attacking.

This bacterial offensive weapon, referred to by microbiologists as the "type III secretory apparatus," provides a clear "survival of the fittest" advantage to organisms that possess it. Presumably, the elements of this structure were duplicated hundreds of millions of years ago, and then recruited for a new use; by combining this with other proteins that had previously been carrying out simpler functions, the entire motor was ultimately generated. Granted, the type III secretory apparatus is just one piece of the flagellum's puzzle, and we are far from filling in the whole picture (if we ever can). But each such new puzzle piece provides a natural explanation for a step that ID had relegated to supernatural forces, and leaves its proponents with smaller and smaller territory to stand upon. Behe cites Darwin's famous sentence to support the arguments of irreducible complexity: "If it could be demonstrated that any complex organ existed, which could not possibly have been formed by numerous, successive, slight modifications, my theory would absolutely break down."[5] In the instance of the flagellum, and in virtually all other instances proposed for irreducible complexity, Darwin's criteria have not been met, and an honest evaluation of current knowledge leads to the same conclusion that follows in Darwin's next sentence: "But I can find out no such case."

THEOLOGICAL OBJECTIONS TO ID

So, scientifically, ID fails to hold up, providing neither an opportunity for experimental validation nor a robust foundation for its primary claim of irreducible complexity. More than that, however, ID also fails in a way that should be more of a concern to the believer than to the hard-nosed scientist. ID is a "God of the gaps" theory, inserting a supposition of the need for supernatural intervention in places that its proponents claim science cannot explain. Various cultures have traditionally tried to ascribe to God various natural phenomena that the science of the day had been unable to sort out—whether a solar eclipse or the beauty of a flower. But those theories have a dismal history. Advances in science ultimately fill in those gaps, to the dismay of those who had attached their faith to them. Ultimately a "God of the gaps" religion runs a huge risk of simply discrediting faith. We must not repeat this mistake in the current era. Intelligent Design fits into this discouraging tradition, and faces the same ultimate demise.

Furthermore, ID portrays the Almighty as a clumsy Creator, having to intervene at regular intervals to fix the inadequacies of His own initial plan for generating the complexity of life. For a believer who stands in awe of the almost unimaginable intelligence and creative genius of God, this is a very unsatisfactory image.

THE FUTURE OF THE ID MOVEMENT

William Dembski, the leading mathematical modeler of the ID movement, deserves credit for emphasizing the overarching importance of seeking out the real truth: "Intelligent Design must not become a noble lie for vanquishing views we find unacceptable (history is full of noble lies that ended in disgrace). Rather, Intelligent Design needs to convince us of its truth on its scientific merits."[6] Dembski is absolutely correct in that assertion, and yet his own statement portends the ultimate demise of ID. Elsewhere, Dembski writes, "If it could be shown that biological systems that are wonderfully complex, elegant and integrated—such as the bacterial flagellum—could have been formed by a gradual Darwinian process (and thus that their specified complexity is an illusion), then Intelligent Design would be refuted on the general grounds that one does not invoke intelligent causes when undirected natural causes will do. In that case, Occam's razor would finish off Intelligent Design quite nicely."[7]

A sober evaluation of current scientific information would have to conclude that this outcome is already at hand. The perceived gaps in evolution that ID intended to fill with God are instead being filled by advances in science. By forcing this limited, narrow view of God's role, Intelligent Design is ironically on a path toward doing considerable damage to faith.

The sincerity of the proponents of Intelligent Design can hardly be questioned. The warm embrace of ID by believers, particularly by evangelical Christians, is completely understandable, given the way in which Darwin's theory has been portrayed by some outspoken evolutionists as demanding atheism. But this ship is not headed to the promised land; it is headed instead to the bottom of the ocean. If believers have attached their last vestiges of hope that God could find a place in human existence through ID theory, and that theory collapses, what then happens to faith?

So is the search for harmony between science and faith hopeless? Must we accept the Dawkins perspective: "The universe we observe has precisely the properties we should expect if there is, at bottom, no design, no purpose, no evil, and no good, nothing but blind pitiless indifference"?[8] May it never be so! To the believer and the scientist alike, I say there is a clear, compelling, and intellectually satisfying solution to this search for the truth.

CHAPTER TEN

Option 4: BioLogos
(Science and Faith in Harmony)

At MY HIGH SCHOOL GRADUATION, an earnest Presbyterian minister, father of one of the graduates, challenged the assembled fidgeting teenagers to consider how they planned to answer life's three great questions: (1) What will be your life's work? (2) What role will love play in your life? and (3) What will you do about faith? The stark directness of his presentation caught all of us by surprise. Being honest with myself, my answers were (1) chemistry; (2) as much as possible; and (3) don't go there. I left the ceremony feeling vaguely uneasy.

A dozen years later I found myself deeply engaged in finding answers to questions 1 and 3. After a long and tortuous path through chemistry, physics, and medicine, I was finally encountering that inspiring field of human endeavor I had been longing to find—one that could combine my love of science and

197

mathematics with a desire to help others—the discipline of medical genetics. At the same time, I had reached the conclusion that faith in God was much more compelling than the atheism I had previously embraced, and I was beginning for the first time in my life to perceive some of the eternal truths of the Bible.

I was vaguely aware that some of those around me thought that this pairing of explorations was contradictory and I was headed over a cliff, but I found it difficult to imagine that there could be a real conflict between scientific truth and spiritual truth. Truth is truth. Truth cannot disprove truth. I joined the American Scientific Affiliation (www.asa3.org), a group of several thousand scientists who are serious believers in God, and found in their meetings and their journal many thoughtful proposals of a pathway toward harmony between science and faith. That was enough for me at that point—to see that other sincere believers were totally comfortable merging their faith with rigorous science.

I confess that I didn't pay much more attention to the potential for conflict between science and faith for several years—it just didn't seem that important. There was too much to discover in scientific research about human genetics, and too much to discover about the nature of God from reading and discussing faith with other believers.

The need to find my own harmony of the worldviews ultimately came as the study of genomes—our own and that of many other organisms on the planet—began to take off, providing an incredibly rich and detailed view of how descent by modification from a common ancestor has occurred. Rather than finding this unsettling, I found this elegant evidence of the relatedness of all living things an occasion of awe, and came to see this as the master plan of the same Almighty who caused the universe to come into being and set its physical parameters just precisely right to allow the creation of stars, planets, heavy elements, and life itself. Without knowing its name at the time, I settled comfortably into a synthesis generally referred to as "theistic evolution," a position I find enormously satisfying to this day.

WHAT IS THEISTIC EVOLUTION?

Mountains of material, in fact entire library shelves, are devoted to the topics of Darwinian evolution, creationism, and Intelligent Design. Yet few scientists or believers are familiar with the term "theistic evolution," sometimes abbreviated "TE." By the now standard criterion of Google search engine entries, there is only one mention of theistic evolution for every ten about creationism and every 140 about Intelligent Design.

Yet theistic evolution is the dominant position of serious biologists who are also serious believers. That includes Asa Gray, Darwin's chief advocate in the United States, and Theodosius Dobzhansky, the twentieth-century architect of evolutionary thinking. It is the view espoused by many Hindus, Muslims, Jews, and Christians, including Pope John Paul II. While it is risky to make presumptions about historical figures, I believe that this is also the view that Maimonides (the highly regarded twelfth-century Jewish philosopher) and Saint Augustine would espouse today if they were presented with the scientific evidence for evolution.

There are many subtle variants of theistic evolution, but a typical version rests upon the following premises:

1. The universe came into being out of nothingness, approximately 14 billion years ago.
2. Despite massive improbabilities, the properties of the universe appear to have been precisely tuned for life.
3. While the precise mechanism of the origin of life on earth remains unknown, once life arose, the process of evolution and natural selection permitted the development of biological diversity and complexity over very long periods of time.
4. Once evolution got under way, no special supernatural intervention was required.
5. Humans are part of this process, sharing a common ancestor with the great apes.
6. But humans are also unique in ways that defy evolutionary explanation and point to our spiritual nature. This includes the existence of the Moral Law (the knowledge of right and wrong) and the search for God that characterizes all human cultures throughout history.

If one accepts these six premises, then an entirely plausible, intellectually satisfying, and logically consistent synthesis emerges: God, who is not limited in space or time, created the universe and established natural laws that govern it. Seeking to populate this otherwise sterile universe with living creatures, God chose the elegant mechanism of evolution to create microbes, plants, and animals of all sorts. Most remarkably, God intentionally chose the same mechanism to give rise to special creatures who would have intelligence, a knowledge of right and wrong, free will, and a desire to seek fellowship with Him. He also knew these creatures would ultimately choose to disobey the Moral Law.

This view is entirely compatible with everything that science teaches us about the natural world. It is also entirely compatible with the great monotheistic religions of the world. The theistic evolution perspective cannot, of course, prove that God is real, as no logical argument can fully achieve that. Belief in God will always require a leap of faith. But this synthesis has provided for legions of scientist-believers a satisfying, consistent, enriching perspective that allows both the scientific and spiritual worldviews to coexist happily within us. This perspective makes it possible for the scientist-believer to be intellectually fulfilled and spiritually alive, both worshiping God and using the tools of science to uncover some of the awesome mysteries of His creation.

CRITIQUES OF THEISTIC EVOLUTION

Of course, many objections to theistic evolution have been raised.[1] If this is such a satisfying synthesis, why is it not more widely embraced? First of all, it is simply not widely known. Few, if any, prominent public advocates have ever spoken passionately about theistic evolution and the way in which it resolves current battles. While many scientists ascribe to TE, they are in general reluctant to speak out for fear of negative reaction from their scientific peers, or perhaps for fear of criticism from the theological community.

On the religious side of the divide, few prominent theologians are currently familiar enough with the details of biological science to endorse this perspective confidently in the face of massive objections from the advocates of creationism or Intelligent Design. Important exceptions can be noted, however. Pope John Paul II in his message to the Pontifical Academy of Sciences in 1996 offered a particularly thoughtful and courageous defense of theistic evolution. The pope stated that "new findings lead us toward the recognition of evolution as more than a hypothesis." He thus accepted the biological reality of evolution, but was careful to balance that with a spiritual perspective, echoing the position of his predecessor Pius XII: "If the origin of the human body comes through living matter which existed previously, the spiritual soul is created directly by God."[2]

This enlightened papal view was warmly welcomed by many believer-scientists. Concerns were raised, however, by comments from Catholic Cardinal Schönborn of Vienna, only months after the death of John Paul II, suggesting that this was a "rather vague and unimportant 1996 letter about evolution," and that more serious consideration should be given to the Intelligent Design perspective.[3] (More recent signals from the Vatican appear to be returning to the perspective of John Paul II.)

Perhaps a more trivial reason that theistic evolution is so little appreciated is that it has a terrible name. Most nontheologians are not quite sure what a theist is, much less how that term could be converted to an adjective and used to modify Darwin's theory. Relegating one's belief in God to an adjective suggests a secondary priority, with the primary emphasis being the noun, namely "evolution." But the alternative of "evolutionary theism" doesn't resonate particularly well either.

Unfortunately, many of the nouns and adjectives that could describe the rich nature of this synthesis are already freighted with so much baggage as to be off-limits. Should we coin the term "crevolution"? Probably not. And one dare not use the words "creation," "intelligent," "fundamental," or "designer," for fear of confusion. We need to start afresh. My modest proposal is to rename theistic evolution as Bios through Logos, or simply BioLogos. Scholars will recognize *bios* as the Greek word for "life" (the root word for biology, biochemistry, and so forth), and *logos* as the Greek for "word." To many believers, the Word is synonymous with God, as powerfully and poetically expressed in those majestic opening lines of the gospel of John, "In the beginning was the Word, and the Word was with God, and the Word was God" (John 1:1). "BioLogos" expresses the belief that God is the source of all life and that life expresses the will of God.

Ironically, another major reason for the invisibility of the BioLogos position is the very harmony that it creates between warring factions. As a society we seem drawn not to harmony but to conflict. The media is partly to blame, but the media only plays to the public's desires. On the evening news you are likely to hear of multicar crackups, destructive hurricanes, violent crimes, messy celebrity divorces, and yes, raucous school board debates over the teaching of evolution. You are not likely to hear much about the coming together of neighborhood groups of different faiths to try to solve community problems, nor about lifelong atheist Anthony Flew becoming a believer, and certainly not about theistic evolution or the double rainbow seen over the city this afternoon. We love conflict and discord, and the harsher the better. In academia, the serious music and art produced by members of the faculty seem to celebrate being hard to listen to and hard to look at. Harmony is boring.

More seriously, however, objections are raised to BioLogos by those who perceive this perspective as doing violence to either science or faith or both. For the atheistic scientist, BioLogos seems to be another "God of the gaps" theory imposing the presence of the divine where none is needed or desired. But this argument is not apt. BioLogos doesn't try to wedge God into gaps in our understanding of the natural world; it proposes God as the answer to questions science was never intended to address, such as "How did the universe get here?" "What is the meaning of life?" "What happens to us after we die?" Unlike Intelligent Design, BioLogos is not intended as a scientific theory. Its truth can be tested only by the spiritual logic of the heart, the mind, and the soul.

The most major current objections to BioLogos arise, however, from believers in God who simply cannot accept that God would have carried out creation using such an apparently random, potentially heartless, and inefficient process as Darwinian evolution. After all, they argue, evolutionists claim that the process is full of chance and random outcomes. If you rewound the clock several hundred million years, and then allowed evolution to proceed forward again, you might end up with a very different outcome. For example, if the now well-documented collision of a large asteroid with the earth 65 million years ago had not happened, it might well be that the emergence of higher intelligence would not have come in the form of a carnivorous mammal (*Homo sapiens*), but in a reptile.

How is this consistent with the theological concept that humans are created "in the image of God" (Genesis 1:27)? Well, perhaps one shouldn't get too hung up on the notion that this scripture is referring to physical anatomy—the image of God seems a lot more about mind than body. Does God have toenails? A belly button?

But how could God take such chances? If evolution is random, how could He really be in charge, and how could He be certain of an outcome that included intelligent beings at all?

The solution is actually readily at hand, once one ceases to apply human limitations to God. If God is outside of nature, then He is outside of space and time. In that context, God could in the moment of creation of the universe also know every detail of the future. That could include the formation of the stars, planets, and galaxies, all of the chemistry, physics, geology, and biology that led to the formation of life on earth, and the evolution of humans, right to the moment of your reading this book—and beyond. In that context, evolution could appear to us to be driven by chance, but from God's perspective the outcome would be entirely specified. Thus, God could be completely and intimately involved in the creation of all species, while from our perspective, limited as it is by the tyranny of linear time, this would appear a random and undirected process.

So perhaps that takes care of the objections about the role of chance in the appearance of humans on this earth. The remaining stumbling block for the BioLogos position, however, at least for most believers, is the apparent conflict of the premises of evolution with important sacred texts. In looking closely at chapters 1 and 2 of the book of Genesis, we have previously concluded that many interpretations have been honorably put forward by sincere believers, and that this powerful document can best be understood as poetry and allegory rather than a literal scientific description of origins. Without repeating those points, consider the words of Theodosius Dobzhansky (1900–1975), a prominent scientist who subscribed to the Russian Orthodox faith and to theistic evolution: "Creation is not an event that happened in 4004 BC; it is a process that began some 10 billion years ago and is still underway. . . . Does the evolutionary doctrine clash with religious faith? It does not. It is a blunder to mistake the Holy Scriptures for elementary textbooks of astronomy, geology, biology, and anthropology. Only if symbols are construed to mean what they are not intended to mean can there arise imaginary, insoluble conflicts."[4]

WHAT ABOUT ADAM AND EVE?

Very well, so the six days of creation can be harmonized with what science tells us about the natural world. But what about the Garden of Eden? Is the description of Adam's creation from the dust of the earth, and the subsequent creation of Eve from one of Adam's ribs, so powerfully described in Genesis 2, a

symbolic allegory of the entrance of the human soul into a previously soulless animal kingdom, or is this intended as literal
history?

As noted previously, studies of human variation, together
with the fossil record, all point to an origin of modern humans
approximately a hundred thousand years ago, most likely in
East Africa. Genetic analyses suggest that approximately ten
thousand ancestors gave rise to the entire population of 6 billion humans on the planet. How, then, does one blend these
scientific observations with the story of Adam and Eve?

In the first place, the biblical texts themselves seem to suggest that there were other humans present at the same time
that Adam and Eve were expelled from the Garden of Eden.
Otherwise, where did Cain's wife, mentioned only after he left
Eden to live in the land of Nod (Genesis 4:16–17), come from?
Some biblical literalists insist that the wives of Cain and Seth
must have been their own sisters, but that is both in serious
conflict with subsequent prohibitions against incest, and incompatible with a straightforward reading of the text. The real
dilemma for the believer comes down to whether Genesis 2 is
describing a special act of miraculous creation that applied to a
historic couple, making them biologically different from all
other creatures that had walked the earth, or whether this is a
poetic and powerful allegory of God's plan for the entrance of
the spiritual nature (the soul) and the Moral Law into humanity.

Since a supernatural God can carry out supernatural acts,
both options are intellectually tenable. However, better minds
than mine have been unable to arrive at a precise understanding of this story over more than three millennia, and so we

should be wary of staking out any position too strongly. Many believers find the story of Adam and Eve compelling as literal history, but no less an intellect than C. S. Lewis, a distinguished scholar of myth and of history, found in the story of Adam and Eve something resembling a moral lesson rather than a scientific textbook or a biography. Here is Lewis's version of the events in question:

> For long centuries, God perfected the animal form which was to become the vehicle of humanity and the image of Himself. He gave it hands whose thumb could be applied to each of the fingers, and jaws and teeth and throat capable of articulation, and a brain sufficiently complex to execute all of the material motions whereby rational thought is incarnated. The creature may have existed in this state for ages before it became man: it may even have been clever enough to make things which a modern archaeologist would accept as proof of its humanity. But it was only an animal because all its physical and psychical processes were directed to purely material and natural ends. Then, in the fullness of time, God caused to descend upon this organism, both on its psychology and physiology, a new kind of consciousness which could say "I" and "me," which could look upon itself as an object, which knew God, which could make judgments of truth, beauty and goodness, and which was so far above time that it could perceive time flowing

past. . . . We do not know how many of these creatures God made, nor how long they continued in
the Paradisal state. But sooner or later they fell.
Someone or something whispered that they could
become as gods. . . . They wanted some corner in
the universe of which they could say to God, "This
is our business, not yours." But there is no such
corner. They wanted to be nouns, but they were,
and eternally must be, mere adjectives. We have no
idea in what particular act, or series of acts, the
self-contradictory, impossible wish found expres-
sion. For all I can see, it might have concerned the
literal eating of a fruit, but the question is of no
consequence.[5]

Conservative Christians who are otherwise great admirers
of C. S. Lewis may be troubled by this passage. Doesn't a compromise on Genesis 1 and 2 start the believer down a slippery
slope, ultimately resulting in the denial of the fundamental
truths of God and His miraculous actions? While there is clear
danger in unrestrained forms of "liberal" theology that eviscer-
ate the real truths of faith, mature observers are used to living
on slippery slopes and deciding where to place a sensible stopping point. Many sacred texts do indeed carry the clear marks
of eyewitness history, and as believers we must hold fast to
those truths. Others, such as the stories of Job and Jonah, and
of Adam and Eve, frankly do not carry that same historical ring.

Given this uncertainty of interpretation of certain scriptural
passages, is it sensible for sincere believers to rest the entirety of

their position in the evolutionary debate, their views on the trust-worthiness of science, and the very foundation of their religious faith on a literalist interpretation, even if other equally sincere be-lievers disagree, and have disagreed even long before Darwin and his *Origin of Species* first appeared? I do not believe that the God who created all the universe, and who communes with His people through prayer and spiritual insight, would expect us to deny the obvious truths of the natural world that science has re-vealed to us, in order to prove our love for Him.

In that context, I find theistic evolution, or BioLogos, to be by far the most scientifically consistent and spiritually satisfying of the alternatives. This position will not go out of style or be disproven by future scientific discoveries. It is intellectually rig-orous, it provides answers to many otherwise puzzling ques-tions, and it allows science and faith to fortify each other like two unshakable pillars, holding up a building called Truth.

SCIENCE AND FAITH: THE CONCLUSION REALLY MATTERS

In the twenty-first century, in an increasingly technological so-ciety, a battle is raging for the hearts and minds of humanity. Many materialists, noting triumphally the advances of science in filling the gaps of our understanding of nature, announce that belief in God is an outmoded superstition, and that we would be better off admitting that and moving on. Many believ-ers in God, convinced that the truth they derive from spiritual introspection is of more enduring value than truths from other sources, see the advances in science and technology as dangerous and untrustworthy. Positions are hardening. Voices are becoming more shrill.

Will we turn our backs on science because it is perceived as a threat to God, abandoning all of the promise of advancing our understanding of nature and applying that to the alleviation of suffering and the betterment of humankind? Alternatively, will we turn our backs on faith, concluding that science has rendered the spiritual life no longer necessary, and that traditional religious symbols can now be replaced by engravings of the double helix on our altars?

Both of these choices are profoundly dangerous. Both deny truth. Both will diminish the nobility of humankind. Both will be devastating to our future. And both are unnecessary. The God of the Bible is also the God of the genome. He can be worshiped in the cathedral or in the laboratory. His creation is majestic, awesome, intricate, and beautiful—and it cannot be at war with itself. Only we imperfect humans can start such battles. And only we can end them.

CHAPTER ELEVEN

Truth Seekers

T HE IMPOVERISHED VILLAGE OF EKU lies in the delta of the Niger River, near the crook in the elbow that makes up the western coastline of Africa. It was there that I learned a powerful and unexpected lesson.

I had traveled to Nigeria in the summer of 1989 to volunteer in a small mission hospital, in order to provide an opportunity for the missionary physicians to attend their annual conference and recharge their spiritual and physical batteries. My college-age daughter and I agreed to go on this adventure together, having long been curious about life in Africa, and having harbored a desire to contribute something to the developing world. I was aware that my own medical skills, dependent as they were upon the high-tech world of an American hospital, might be poorly matched to the challenges of unfamiliar tropical diseases and little technical support. Nonetheless, I arrived in Nigeria with an expectation that my presence there was going to make a significant difference in the lives of the many I expected to care for.

The hospital at Eku was unlike anything I had experienced. There were never enough beds, so patients often had to sleep on the floor. Their families often traveled with them and took on the responsibility of feeding them, since the hospital was not able to provide adequate nourishment. A wide spectrum of severe diseases was represented. Oftentimes patients arrived at the hospital only after many days of progressive illness. Even worse, the course of disease was regularly compounded by the toxic ministrations of the witch doctors, to which many Nigerians would first go for help, coming to the hospital in Eku only when all else failed. Hardest of all for me to accept, it became abundantly clear that the majority of the diseases I was called upon to treat represented a devastating failure of the public health system. Tuberculosis, malaria, tetanus, and a wide variety of parasitic diseases all reflected an environment that was completely unregulated and a health care system that was completely broken.

Overwhelmed by the enormity of these problems, exhausted by the constant stream of patients with illnesses I was poorly equipped to diagnose, frustrated by the lack of laboratory and X-ray support, I grew more and more discouraged, wondering why I had ever thought that this trip would be a good thing.

Then one afternoon in the clinic a young farmer was brought in by his family with progressive weakness and massive swelling of his legs. Taking his pulse, I was startled to note that it essentially disappeared every time he took in a breath. Though I had never seen this classic physical sign (referred to as a "paradoxical pulse") so dramatically demonstrated, I was pretty sure this must mean that this young farmer had accumulated a large amount of fluid in the pericardial sac around his heart. This fluid was threatening to choke off his circulation and take his life.

In this setting, the most likely cause was tuberculosis. We had drugs at Eku for tuberculosis, but they could not act quickly enough to save this young man. He had at most a few days to live unless something drastic was done. The only chance to save him was to carry out a highly risky procedure of drawing off the pericardial fluid with a large bore needle placed in his chest. In the developed world, such a procedure would be done only by a highly trained interventional cardiologist, guided by an ultrasound machine, in order to avoid lacerating the heart and causing immediate death.

No ultrasound was available. No other physician present in this small Nigerian hospital had ever undertaken this procedure. The choice was for me to attempt a highly risky and invasive needle aspiration or watch the farmer die. I explained the situation to the young man, who was now fully aware of his own precarious state. He calmly urged me to proceed. With my heart in my mouth and a prayer on my lips, I inserted a large needle just under his sternum and aimed for his left shoulder, all the while fearing that I might have made the wrong diagnosis, in which case I was almost certainly going to kill him.

I didn't have to wait long. The rush of dark red fluid in my syringe initially made me panic that I might have entered the heart chamber, but it soon became apparent that this was not normal heart's blood. It was a massive amount of bloody tuberculous effusion from the pericardial sac around the heart.

Nearly a quart of fluid was drawn off. The young man's response was dramatic. His paradoxical pulse disappeared almost at once, and within the next twenty-four hours the swelling of his legs rapidly improved.

For a few hours after this experience I felt a great sense of relief, even elation, at what had happened. But by the next morning, the same familiar gloom began to settle over me. After all, the circumstances that had led this young man to acquire tuberculosis were not going to change. He would be started on TB drugs in the hospital, yet the chances were good that he would not have the resources to pay for the entire two years of treatment that he needed, and he might very well suffer a recurrence and die despite our efforts. Even if he survived the disease, some other preventable disorder, born of dirty water, inadequate nutrition, and a dangerous environment, probably lay not too far in his future. The chances for long life in a Nigerian farmer are poor.

With those discouraging thoughts in my head, I approached his bedside the next morning, finding him reading his Bible. He looked at me quizzically, and asked whether I had worked at the hospital for a long time. I admitted that I was new, feeling somewhat irritated and embarrassed that it had been so easy for him to figure that out. But then this young Nigerian farmer, just about as different from me in culture, experience, and ancestry as any two humans could be, spoke the words that will forever be emblazoned in my mind: "I get the sense you are wondering why you came here," he said. "I have an answer for you. You came here for one reason. You came here for me."

I was stunned. Stunned that he could see so clearly into my heart, but even more stunned at the words he was speaking. I had plunged a needle close to his heart; he had directly impaled mine. With a few simple words he had put my grandiose dreams of being the great white doctor, healing the African millions, to shame. He was right. We are each called to reach out to others. On rare occasions that can happen on a grand scale. But most of the time it happens in simple acts of kindness of one person to another. Those are the events that really matter. The tears of relief that blurred my vision as I digested his words stemmed from indescribable reassurance—reassurance that there in that strange place for just that one moment, I was in harmony with God's will, bonded together with this young man in a most unlikely but marvelous way.

Nothing I had learned from science could explain that experience. Nothing about the evolutionary explanations for human behavior could account for why it seemed so right for this privileged white man to be standing at the bedside of this young African farmer, each of them receiving something exceptional. This was what C. S. Lewis calls agape. It is the love that seeks no recompense. It is an affront to materialism and naturalism. And it is the sweetest joy that one can experience.

In years of dreaming of going to Africa, I had felt the gentle stirrings of a desire to do something truly unselfish for others— that calling to serve with no expectation of personal benefit that is common to all human cultures. But I had let other, less noble dreams get in the way—the expectation of receiving admiration from the Eku villagers, the anticipation of applause from my medical colleagues at home. Those grand schemes were clearly not happening for me in the gritty reality of impoverished Eku. But the simple act of trying to help just one person, in a desperate situation where my skills were poorly matched to the challenge, turned out to represent the most meaningful of all human experiences. A burden lifted. This was true north. And the compass pointed not at self-glorification, or at materialism, or even at medical science—instead it pointed at the goodness that we all hope desperately to find within ourselves and others. I also saw more clearly than ever before the author of that goodness and truth, the real True North, God himself, revealing His holy nature by the way in which He has written this desire to seek goodness in all of our hearts.

Making Personal Sense of the Evidence

So here, in the final chapter, we have come full circle, returning again to the existence of the Moral Law, where this story began. We have traveled through the sciences of chemistry, physics, cosmology, geology, paleontology, and biology—and yet this uniquely human attribute still causes wonder. After twenty-eight years as a believer, the Moral Law still stands out for me as the strongest signpost to God. More than that, it points to a God who cares about human beings, and a God who is infinitely good and holy.

The other observations, discussed earlier, that point to a

Creator—the fact that the universe had a beginning, that it obeys orderly laws that can be expressed precisely with mathematics, and the existence of a remarkable series of "coincidences" that allow the laws of nature to support life—do not tell us much about what kind of God must be behind it all, but they do point toward an intelligent mind that could lie behind such precise and elegant principles. But what kind of mind? What, exactly, should we believe?

WHAT KIND OF FAITH?

In the opening chapter of this book, I described my own pathway from atheism to belief. I now owe you a deeper explanation of my subsequent path. I offer this with some trepidation, since strong passions tend to be incited as soon as one begins to differentiate from a general sense of God's existence to a specific set of beliefs.

Most of the world's great faiths share many truths, and probably they would not have survived had that not been so. Yet there are also interesting and important differences, and each person needs to seek out his own particular path to the truth.

After my conversion to belief in God, I spent considerable time trying to discern His characteristics. I concluded that He must be a God who cares about persons, or the argument about the Moral Law would not make much sense. So deism wouldn't do for me. I also concluded that God must be holy and righteous, since the Moral Law calls me in that direction. But this still seemed awfully abstract. Just because God is good and loves His creatures does not, for instance, require that we have the ability to communicate with Him, or to have some sort of relationship with Him. I found an increasing sense of longing for that, however, and I began to realize that this is what prayer is all about. Prayer is not, as some seem to suggest, an opportunity to manipulate God into doing what you want Him to. Prayer is instead our way of seeking fellowship with God, learning about Him, and attempting to perceive His perspective on the many issues around us that cause us puzzlement, wonder, or distress.

Yet I found it difficult to build that bridge toward God. The more I learned about Him, the more His purity and holiness seemed unapproachable, and the darker my own thoughts and actions seemed to be in that bright light.

I began to be increasingly aware of my own inability to do the right thing, even for a day. I could generate lots of excuses, but when I was really honest with myself, pride, apathy, and anger were regularly winning my internal battles. I had never really thought of applying the word "sinner" to myself before, but now it was painfully obvious that this old-fashioned word, one from which I had previously recoiled because it seemed coarse and judgmental, fit quite accurately.

I sought to engineer a cure by spending more time in self-examination and prayer. But those efforts proved largely dry and unrewarding, failing to carry me across the widening gap between my awareness of my imperfect nature and God's perfection.

Into this deepening gloom came the person of Jesus Christ.

During my boyhood years sitting in the choir loft of a Christian church, I really had no idea who Christ was. I thought of Him as a myth, a fairy tale, a superhero in a "just so" bedtime story. But as I read the actual account of His life for the first time in the four gospels, the eyewitness nature of the narratives and the enormity of Christ's claims and their consequences gradually began to sink in. Here was a man who not only claimed to *know* God, He claimed to *be* God. No other figure I could find in any other faith made such an outrageous claim. He also claimed to be able to forgive sins, which seemed both exciting and utterly shocking. He was humble and loving, He spoke remarkable words of wisdom, and yet He was put to death on the cross by those who feared Him. He was a man, so He knew the human condition that I was finding so burdensome, and yet He promised to relieve that burden: "Come unto me all ye that are weary and burdened, and I will give you rest" (Matthew 11:28).

The other scandalous thing that the New Testament eyewitnesses said about Him, and that Christians seemed to take as a central tenet of their faith, is that this good man rose from the dead. For a scientific mind, this was difficult stuff. But on the other hand, if Christ really was the Son of God, as He explicitly claimed, then surely of all those who had ever walked the earth, He could suspend the laws of nature if He needed to do so to achieve a more important purpose.

But His resurrection had to be more than a demonstration of magical powers. What was the real point of it? Christians have puzzled over this question for two millennia. After much searching, I could find no single answer—instead, there were several interlocking answers, all pointing to the idea of a bridge between our sinful selves and a holy God. Some commentators focus on the idea of substitution—Christ dying in the place of all of us who deserve God's judgment for our wrongdoings. Others call it redemption—Christ paid the ultimate price to free us from the bondage of sin, so that we could find God and rest in the confidence that He no longer judges us by our actions, but sees us as having been washed clean. Christians call this salvation by grace. But for me, the crucifixion and resurrection also provided something else. My desire to draw close to God was blocked by my own pride and sinfulness, which in turn was an inevitable consequence of my own selfish desire to be in control. Faithfulness to God required a kind of death of self-will, in order to be reborn as a new creation.

How could I achieve such a thing? As had happened so many times with previous dilemmas, the words of C. S. Lewis captured the answer precisely:

> But supposing God became a man—suppose our human nature which can suffer and die was amalgamated with God's nature in one person—then that person could help us. He could surrender His will, and suffer and die, because He was man; and He could do it perfectly because He was God. You and I can go through this process only if God does it in us; but God can do it only if He becomes man. Our attempts at this dying will succeed only if we men share in God's dying, just as our thinking can succeed only because it is a drop out of the ocean of His intelligence: but we cannot share God's dying

> unless God dies; and He cannot die except by being
> a man. That is the sense in which He pays our debt,
> and suffers for us what He Himself need not suffer
> at all.[1]

Before I became a believer in God, this kind of logic seemed like utter nonsense. Now the crucifixion and resurrection emerged as the compelling solution to the gap that yawned between God and myself, a gap that could now be bridged by the person of Jesus Christ.

So I became convinced that God's arrival on earth in the person of Jesus Christ could serve a divine purpose. But did this mesh with history? The scientist in me refused to go any further along this path toward Christian belief, no matter how appealing, if the biblical writings about Christ turned out to be a myth or, worse yet, a hoax. But the more I read of biblical and non-biblical accounts of events in first-century Palestine, the more amazed I was at the historical evidence for the existence of Jesus Christ. First of all, the gospels of Matthew, Mark, Luke, and John were put down just a few decades after Christ's death. Their style and content suggests strongly that they are intended to be the record of eyewitnesses (Matthew and John were among the twelve apostles). Concerns about errors creeping in by successive copying or bad translation have been mostly laid to rest by discovery of very early manuscripts. Thus, the evidence for authenticity of the four gospels turns out to be quite strong. Furthermore, non-Christian historians of the first century such as Josephus bear witness to a Jewish prophet who was crucified by Pontius Pilate around 33 A.D. Many more examples of evidence for the historical nature of Christ's existence have been collected in many excellent books, to which the interested reader is referred.[2] In fact, one scholar has written, "The historicity of Christ is as axiomatic for an unbiased historian as the historicity of Julius Caesar."[3]

EVIDENCE DEMANDING A VERDICT

So the growing evidence of this unique individual, who seemed to represent God in search of man (whereas most other religions seemed to be man in search of God) provided a compelling case. But I hesitated, afraid of the consequences, and afflicted by doubts. Maybe Christ was just a great spiritual teacher? Again, Lewis seemed to have written one particular paragraph just for me:

> I am trying here to prevent anyone saying the really
> foolish thing that people often say about Him: "I'm
> ready to accept Jesus as a great moral teacher, but I
> don't accept His claim to be God." That is one thing
> we must not say. A man who was merely a man
> and said the sort of things Jesus said would not be a
> great moral teacher. He would either be a lunatic—
> on a level with a man who says He is a poached
> egg—or else He would be the Devil of Hell. You
> must make your choice. Either this man was, and
> is, the Son of God: or else a madman or something
> worse. You can shut Him up for a fool, you can spit

at Him and kill Him as a demon; or you can fall at His feet and call Him Lord and God. But let us not come with any patronizing nonsense about His being a great human teacher. He has not left that open to us. He did not intend to.[4]

Lewis was right. I had to make a choice. A full year had passed since I decided to believe in some sort of God, and now I was being called to account. On a beautiful fall day, as I was hiking in the Cascade Mountains during my first trip west of the Mississippi, the majesty and beauty of God's creation overwhelmed my resistance. As I rounded a corner and saw a beautiful and unexpected frozen waterfall, hundreds of feet high, I knew the search was over. The next morning, I knelt in the dewy grass as the sun rose and surrendered to Jesus Christ.

I do not mean by telling this story to evangelize or proselytize. Each person must carry out his or her own search for spiritual truth. If God is real, He will assist. Far too much has been said by Christians about the exclusive club they inhabit. Tolerance is a virtue; intolerance is a vice. I find it deeply disturbing when believers in one faith tradition dismiss the spiritual experiences of others. Regrettably, Christians seem particularly prone to do this. Personally, I have found much to learn from and admire in other spiritual traditions, though I have found the special revelation of God's nature in Jesus Christ to be an essential component of my own faith.

Christians all too often come across as arrogant, judgmental, and self-righteous, but Christ never did. Consider, for instance, the well-known parable of the Good Samaritan. The nature of the participants in this morality play would have been immediately apparent to listeners in Christ's day, though less so in modern times. Here are Jesus' words, as recorded in Luke 10:30–37:

> A man was going down from Jerusalem to Jericho, when he fell into the hands of robbers. They stripped him of his clothes, beat him and went away, leaving him half dead. A priest happened to be going down the same road, and when he saw the man, he passed by on the other side. So, too, a Levite, when he came to the place and saw him, passed by on the other side. But a Samaritan, as he traveled, came where the man was; and when he saw him, he took pity on him. He went to him and bandaged his wounds, pouring on oil and wine. Then he put the man on his own donkey, took him to an inn and took care of him. The next day he took out two silver coins and gave them to the innkeeper. "Look after him," he said, "and when I return, I will reimburse you for any extra expense you may have." Which of these three do you think was the neighbor to the man who fell into the hands of robbers? The expert in the law replied, "The one who had mercy on him." Jesus told him, "Go and do likewise."

Samaritans were much hated by the Jews, because they rejected many of the teachings of the Jewish prophets. The fact that Jesus would put forward the behavior of the Samaritan as more virtuous than that of a priest or a lay leader (a Levite) must have been scandalous to his hearers. But the overarching principle of love and acceptance appears throughout Christ's teachings in the New Testament. It is the most important guide of how we are to treat others. In Matthew 22:35 Christ is queried about which is the greatest of God's commandments. He answers simply, "Love the Lord your God with all your heart and with all your soul and with all your mind. This is the first and greatest commandment. And the second is like it: Love your neighbor as yourself."

Many of these principles can be found in other great religions of the world. Yet if faith is not just a cultural practice, but rather a search for absolute truth, we must not go so far as to commit the logical fallacy of saying that all conflicting points of view are equally true. Monotheism and polytheism cannot both be right. Through my own search, Christianity has provided for me that special ring of eternal truth. But you must conduct your own search.

Seek and Ye Shall Find

If you have made it this far with me, I hope you will agree that the scientific and spiritual worldviews both have much to offer. Both provide differing but complementary ways of answering the greatest of the world's questions, and both can coexist happily within the mind of an intellectually inquisitive person living in the twenty-first century.

Science is the only legitimate way to investigate the natural world. Whether probing the structure of the atom, the nature of the cosmos, or the DNA sequence of the human genome, the scientific method is the only reliable way to seek out the truth of natural events. Yes, experiments can fail spectacularly, interpretations of experiments can be misguided, and science can make mistakes. But the nature of science is self-correcting. No major fallacy can long persist in the face of a progressive increase in knowledge.

Nevertheless, science alone is not enough to answer all the important questions. Even Albert Einstein saw the poverty of a purely naturalistic worldview. Choosing his words carefully, he wrote, "Science without religion is lame, religion without science is blind."[5] The meaning of human existence, the reality of God, the possibility of an afterlife, and many other spiritual questions lie outside of the reach of the scientific method. While an atheist may claim that those questions are therefore unanswerable and irrelevant, that does not resonate with most individuals' human experience. John Polkinghorne argues this point cogently by a comparison to music:

> The poverty of an objectivistic account is made only too clear when we consider the mystery of music. From a scientific point of view, it is nothing but vibrations in the air, impinging on the eardrums and stimulating neural currents in the brain.
>
> How does it come about that this banal sequence of temporal activity has the power to speak to our hearts of an eternal beauty? The whole range

of subjective experience, from perceiving a patch of pink, to being enthralled by a performance of the Mass in B Minor, and on to the mystic's encounter with the ineffable reality of the One, all these truly human experiences are at the center of our encounter with reality, and they are not to be dismissed as epiphenomenal froth on the surface of a universe whose true nature is impersonal and lifeless.[6]

Science is not the only way of knowing. The spiritual worldview provides another way of finding truth. Scientists who deny this would be well advised to consider the limits of their own tools, as nicely represented in a parable told by the astronomer Arthur Eddington. He described a man who set about to study deep-sea life using a net that had a mesh size of three inches. After catching many wild and wonderful creatures from the depths, the man concluded that there are no deep-sea fish that are smaller than three inches in length! If we are using the scientific net to catch our particular version of truth, we should not be surprised that it does not catch the evidence of spirit.

What obstacles lie in the way of a broader embrace of the complementary nature of the scientific and spiritual worldviews? This is not just a theoretical question for dry philosophical consideration. It is a challenge for each one of us. I hope you will forgive me, therefore, if I address you somewhat more personally as we approach the end of this book.

AN EXHORTATION TO BELIEVERS

If you are a believer in God who picked up this book because of concerns that science is eroding faith by promoting an atheistic worldview, I hope you are reassured by the potential for harmony between faith and science. If God is the Creator of all the universe, if God had a specific plan for the arrival of humankind on the scene, and if He had a desire for personal fellowship with humans, into whom He had instilled the Moral Law as a signpost toward Himself, then He can hardly be threatened by the efforts of our puny minds to understand the grandeur of His creation.

In that context, science can be a form of worship. Indeed, believers should seek to be in the forefront among those chasing after new knowledge. Believers have led science at many times in the past. Yet all too often today, scientists are uneasy about admitting their spiritual views. To add to the problem, church leaders often seem to be out of step with new scientific findings, and run the risk of attacking scientific perspectives without fully understanding the facts. The consequence can bring ridicule on the church, driving sincere seekers away from God instead of into His arms. Proverbs 19:2 warns against this kind of well-intentioned but misinformed religious fervor: "It is not good to have zeal without knowledge."

Believers would do well to follow the exhortation of Copernicus, who found in the discovery that the earth revolves around the sun an opportunity to celebrate, rather than diminish, the grandeur of God: "To know the mighty works of God; to comprehend His wisdom and majesty and power; to appreciate, in degree, the wonderful working of His laws, surely all this must be a pleasing and acceptable mode of worship to the Most High, to whom ignorance cannot be more grateful than knowledge."[7]

AN EXHORTATION TO SCIENTISTS

On the other hand, if you are one who trusts the methods of science but remains skeptical about faith, this would be a good moment to ask yourself what barriers lie in your way toward seeking a harmony between these worldviews.

Have you been concerned that belief in God requires a descent into irrationality, a compromise of logic, or even intellectual suicide? It is hoped that the arguments presented within this book will provide at least a partial antidote to that view, and will convince you that of all the possible worldviews, atheism is the least rational.

Have you been turned off by the hypocritical behavior of those who profess belief? Again, keep in mind that the pure water of spiritual truth is carried in those rusty containers called human beings, so there should be no surprise that at times those foundational beliefs can be severely distorted. Do not rest your evaluation of faith, therefore, on what you see in the behavior of individual humans or of organized religion. Rest it instead on the timeless spiritual truths that faith presents.

Are you distressed by some specific philosophical problem with faith, such as why a loving God would allow suffering? Recognize that a great deal of suffering is brought upon us by our own actions or those of others, and that in a world where humans practice free will, it is inevitable. Understand, also, that if God is real, His purposes will often not be the same as ours. Hard though it is to accept, a complete absence of suffering may not be in the best interest of our spiritual growth.

Are you simply uncomfortable accepting the idea that the tools of science are insufficient for answering any important question? This is particularly a problem for scientists, who have committed their lives to the experimental assessment of reality. From that perspective, admitting the inability of science to answer all questions can be a blow to our intellectual pride—but that blow needs to be recognized, internalized, and learned from.

Does this discussion of spirituality simply make you uncomfortable, because of a sense that recognizing the possibility of God might place new requirements on your own life plans and actions? I recognize this reaction clearly from my own period of "willful blindness," and yet I can testify that coming to a knowledge of God's love and grace is empowering, not constraining. God is in the business of release, not incarceration.

And finally, have you simply not taken the time to seriously consider the spiritual worldview? In our modern world, too many of us are rushing from experience to experience, trying to deny our own mortality, and putting off any serious consideration of God until some future moment when we imagine the circumstances will be right.

Life is short. The death rate will be one per person for the foreseeable future. Opening one's self to the life of the spirit can be indescribably enriching. Don't put off a consideration of these questions of eternal significance until some personal crisis or advancing age forces a recognition of spiritual impoverishment.

A FINAL WORD

Seekers, there are answers to these questions. There is joy and peace to be found in the harmony of God's creation. In the upstairs hall of my home hangs a beautifully decorated pair of scripture verses, illuminated in many colors by the hand of my daughter. I come back to those verses many times when I am struggling for answers, and they never fail to remind me of the nature of true wisdom: "But if any of you lacks wisdom, let him ask of God, who gives to all men generously and without reproach, and it will be given him" (James 1:5). "The wisdom from above is first pure, then peaceable, gentle, reasonable, full of mercy and good fruits, unwavering, without hypocrisy" (James 3:17).

My prayer for our hurting world is that we would together, with love, understanding, and compassion, seek and find that kind of wisdom.

It is time to call a truce in the escalating war between science and spirit. The war was never really necessary. Like so many earthly wars, this one has been initiated and intensified by extremists on both sides, sounding alarms that predict imminent ruin unless the other side is vanquished. Science is not threatened by God; it is enhanced. God is most certainly not threatened by science; He made it all possible. So let us together seek to reclaim the solid ground of an intellectually and spiritually satisfying synthesis of *all* great truths. That ancient motherland of reason and worship was never in danger of crumbling. It never will be. It beckons all sincere seekers of truth to come and take up residence there. Answer that call. Abandon the battlements. Our hopes, joys, and the future of our world depend on it.

The Moral Practice of Science and Medicine: Bioethics

MANY MEMBERS OF THE GENERAL PUBLIC are excited about the potential of advances in biomedical research to prevent or cure terrible diseases, but are also anxious about whether these new technologies are leading us into dangerous territory. The discipline that considers the morality of applications of biotechnology and medicine to humanity is called bioethics. In this Appendix, we will consider a sample of some of the bioethical dilemmas that are inspiring significant debate today—though this is by no means an exhaustive list. I will focus particularly on advances that are arising from the rapid progress in understanding the human genome.

MEDICAL GENETICS

Some years ago, a young woman came to the oncology clinic at the University of Michigan on a desperate mission. It was the day that I realized a real revolution in genetic medicine was beginning. She and I were brought together by a tangled set of circumstances involving a close-knit family, a terrible disease, and the leading edge of research on the human genome.[1]

Susan (not her real name) and her family lived under a cloud. First her mother had been diagnosed with breast cancer, then her aunt, then two of her aunt's children, and then Susan's oldest sister. Deeply alarmed, Susan was careful to examine herself and obtain regular mammograms, while watching her sister ultimately lose her battle. One of Susan's cousins elected to undergo a prophylactic double mastectomy, in hopes of avoiding the same fate. Then Susan's remaining sister, Janet, found a lump, and it too proved to be cancer.

Meanwhile, my physician colleague Barbara Weber and I had initiated a project in Michigan to try to identify hereditary factors in breast cancer. Susan's family enrolled in the study, and were known to me only as "Family 15." But by one of those strange coincidences, when Janet came for counseling about her new diagnosis of breast cancer, it was Dr. Weber who saw her in the clinic, heard about the family history, and realized the connection.

Susan's desperate mission a few months later was to see whether Dr. Weber and I had any further information from the research study that would dissuade her from proceeding with a double mastectomy. No longer able to be optimistic, she had scheduled this drastic procedure to take place in three days. The timing of her visit was exquisite. Work done in our laboratory over the preceding weeks had demonstrated that there was an extremely high likelihood that members of Susan's family were in fact carrying a dangerous mutation in a gene (now known as BRCA1) on chromosome 17. We had started the study with little expectation that such important clinical applications could occur so quickly. Now, however, faced with an urgent situation, Dr. Weber and I agreed it would be unethical to withhold information at a time where it had such obvious relevance.

Going back to the lab and poring over the data made it immediately clear that Susan did not inherit the dangerous mutation that her mother and her two sisters carried, and therefore her risk of breast cancer was no higher than the average woman's. On that day, Susan became the first person on earth to receive information about her BRCA1 status. Her reaction was a mix of elation and disbelief. She canceled the surgery.

Word spread through her family like wildfire, and the phone began ringing off the hook. Within a few weeks, Dr. Weber and I found ourselves counseling her large extended family, all of whom wished to know their status.

There were many additional dramatic moments. The cousin who had had the double mastectomy years before turned out not to carry the dangerous mutation after all. Initially stunned when told of this result, she ultimately came to peace with it, concluding that she had made the best choice she could at the time that she had decided to have the surgery.

Perhaps most dramatic were the consequences for another branch of the family who had previously thought that they were at no increased risk for breast cancer, since they were related through their father to the affected women. The idea that a susceptibility gene for breast cancer could be transmitted by unaffected males had not seemed plausible, but that's how the BRCA1 gene works. In fact, it turned out that their father carried the mutation and had passed this on to five of his ten children. One of them, aged thirty-nine, was stunned by the news that she might be at risk. She wanted to know her DNA result; it was positive. She immediately asked for a mammogram to be performed, and that same day learned that she had breast cancer. The good news was that the tumor was very small, and would probably not otherwise have been diagnosed for another two or three years, at which point the prognosis might not have been nearly so encouraging.

All told, thirty-five members of this single family were found to be at risk. About half of them turned out to carry the dangerous mutation, and half of those were women. Women who carry this gene are at risk for both breast and ovarian cancer. The medical and psychological consequences have been profound. Even Susan, who escaped "the curse," went through a prolonged period of depression and a sense of alienation from her family, experiencing what is known as "survivor guilt," named after those who lived through the Holocaust.

Susan's family was admittedly unusual. Most breast cancer has hereditary contributions, but not nearly as strong as in her family. But there are no perfect specimens among us. The universal presence of mutations in DNA, the price we pay for evolution, means that no one can claim bodily perfection any more than spiritual perfection.

The time is coming soon when the genetic glitches that place each of us at risk for some future illness will be discovered, and we may each have an opportunity, as Susan's family did, to find out what's hiding within our own DNA instruction book. As we begin to look at the consequences of these rapid advances in understanding human biology, ethical questions arise, and well they should. Knowledge itself has no intrinsic moral value; it is the way in which that knowledge is put to use that acquires an ethical dimension. This principle should be familiar from many nonmedical applications in daily experience. For example, certain mixtures of chemicals can generate a colorful fireworks display that will brighten our skies and lift our spirits at a time of celebration. The same mixture can be used, however, to fire a projectile, or make a bomb that kills dozens of innocent civilians.

There are compelling reasons to celebrate the outpouring of scientific advances that arise from the Human Genome Project. After all, in virtually every culture throughout history, the alleviation of suffering from medical illness has been considered a good thing, perhaps even an ethical mandate. Thus, while some might argue that science is moving too quickly, and that we should declare a moratorium on certain applications until we have time to study them ethically, I find those arguments difficult to convey to parents who are desperate to help an ailing child. Would not intentional restrictions on the progress of life-saving science, simply to allow ethics to "catch up," be themselves unethical?

PERSONALIZED MEDICINE

What can one expect in the coming years from the current revolution in genomics? First of all, the understanding of that tiny fraction (0.1 percent) of the human DNA that differs from person to person has moved ahead rapidly, and is likely in the next few years to reveal the most common genetic glitches that place individuals at risk for cancer, diabetes, heart disease, Alzheimer's disease, and many other conditions. It will allow each of us, if we're interested, to obtain a personal readout documenting our future risks of illness. Few of those reports will be as dramatic as in Susan's family, however, because few of us will have genetic glitches with such strong effects. Would you want to know? Many people would say yes, if interventions are available to reduce their risks, and in some instances, that is already possible. A person found to be at high genetic risk for colon cancer, for instance, can begin colonoscopy at an early age, and repeat it faithfully once a year in order to detect the small polyps at a time when they can be readily removed, preventing an ultimate transformation into a deadly cancer. Individuals found to be at higher than average risk for diabetes can watch their diet carefully and avoid gaining weight. Individuals at high risk for blood clots in the legs can avoid birth control pills and prolonged periods of immobilization.

In another powerful application of personalized medicine, it is increasingly clear that an individual's response to drugs is heavily influenced by heredity. It may be possible in many instances to predict who should be given which drug, and at what dose, by first testing a DNA sample from that person. This "pharmacogenomics" approach, applied widely, should result in increasingly effective drug therapy, and fewer occurrences of dangerous or even fatal side effects.

Ethical Problems Posed by DNA Testing

The advances described above are all potentially valuable. Yet many ethical dilemmas have also been encountered. In Susan's family, a strong disagreement arose about whether it was appropriate to test children for the presence of a BRCA1 mutation. Since no medical intervention was available for children, and since the psychological impact of positive testing could be substantial, Dr. Weber and I, supported by a majority of ethical experts we consulted, concluded that such testing should be delayed until the individual reached the age of eighteen. In at least one instance, a father who carried the BRCA1 mutation became quite angry that his daughters could not be tested right then. He argued that his parental authority ought to trump our decision.

A larger ethical debate has arisen over whether or not it is ever appropriate for third parties to have access to or to use genetic information about individuals. Susan and many of her relatives were fearful that if they tested positive, information might fall into the hands of their health insurance companies, or their employers, and they might find themselves without medical coverage or a job.

Extensive ethical analysis of this situation has led to the conclusion that such discriminatory use of genetic information would be a violation of principles of justice and fairness, since flaws in DNA are essentially universal, and no one gets to pick his own DNA sequence. On the other hand, if insurance customers know their own risks, but insurers do not, then there is a risk that the customers will game the system. That could be a significant issue for large life insurance policies. It does not appear to play much of a role at all in health insurance.

The weight of evidence suggests, therefore, that legislative protection ought to be provided against genetic discrimination in health insurance and the workplace. At this writing, however, we still await the implementation of effective legislation at the federal level in the United States. Failure to provide legal protection could have a profoundly negative effect on the future of individualized preventive medicine, since individuals may be frightened to obtain genetic information that otherwise could be quite useful to them.

Another major ethical question that arises in these discussions, and rightfully so, is the issue of access to care. This is particularly vexing in the United States, where at this writing more than 40 million of its citizens lack health insurance coverage. Of all the developed nations in the world, we in the United States seem most able to turn our heads and look away from this failure of moral responsibility. One of the tragic consequences is relegating the impoverished to highly inefficient and inconsistent emergency room care. This does nothing for prevention, focusing mainly on the medical disasters when they inevitably occur.

The access dilemma will become ever more acute as advances in research, particularly inspired by what we are learning about the genome, lead to new and much more effective means of prevention for cancer, heart disease, mental illness, and many other conditions.

BIOETHICS RESTS ON THE FOUNDATION OF THE MORAL LAW

Before delving further into ethical dilemmas, it behooves us to consider the foundations upon which our judgments of ethical behavior are based. Many bioethical issues are complicated. Those debating the morality of a given decision may come from vastly different cultural backgrounds and religious traditions. In a secular and pluralistic society, is it realistic that any group could agree on the right course of action in difficult circumstances?

Actually, I have found that once the facts of the matter are clear, in most instances people with widely different world-views can come to a comfortable and shared conclusion. While that may at first seem surprising, I believe that it is a compelling example of the universality of the Moral Law. We all have an innate knowledge of right and wrong; although that can be obscured by distractions and misunderstandings, it can also be discovered through careful contemplation. T. L. Beauchamp and J. F. Childress[2] argue that four ethical principles undergird much of bioethics, and are common to virtually all cultures and societies. These include

1. *Respect for autonomy*—the principle that a rational individual should be given freedom in personal decision making, without undue outside coercion

2. *Justice*—the requirement for fair, moral, and impartial treatment of all persons
3. *Beneficence*—the mandate to treat others in their best interest
4. *Nonmaleficence*—"First do no harm" (as in the Hippocratic Oath)

WHAT ROLE SHOULD FAITH PLAY IN BIOETHICAL DEBATES?

A religious person will see these as principles clearly laid out in sacred texts of the Judaeo-Christian, Islamic, Buddhist, and other religious traditions. In fact, some of the most eloquent and powerful statements of these principles are to be found in such sacred texts. But one need not be a theist to agree to these principles. Even a person untrained in musical theory can be transported by a Mozart concerto. The Moral Law speaks to all of us, whether or not we agree on its origins.

Basic principles of ethics can be derived from the Moral Law, and are universal. But conflicts can arise in a situation where not all of the principles can be satisfied at the same time, and different observers attach different weights to the principles that must be somehow balanced. In many instances, society has reached a consensus on how to handle this; in other instances, such as the one we are about to consider next, reasonable people will disagree about the ethical balance sheet.

STEM CELLS AND CLONING

I still recall the Sunday afternoon several years ago when a reporter called me at home to seek my opinion about a paper about to be published in a prominent journal, reporting the cloning of Dolly the sheep. This was an astounding and unprecedented development, as virtually all scientists (including me) thought that it would be impossible to clone a mammal. Although the entire DNA instruction book of an organism is carried in each cell of the body, it was assumed that irreversible changes in that DNA would make it impossible for an accurate and complete instruction book to be reprogrammed in this way.

We were wrong. Indeed, over the course of the last decade, discovery after discovery is revealing the remarkable and completely unanticipated plasticity of mammalian cell types. That in turn has led to the current controversy about the potential benefits and risks of this kind of research, characterized by intense public disagreements that show no sign of lessening.

The debates about human stem cells, in particular, have been so heated, and the jargon has been so impenetrable, that a bit of background is necessary. A stem cell is one that carries within it the potential to develop into several different types of cells. In the bone marrow, for instance, a stem cell can give rise to red blood cells, white blood cells, bone cells, and even, given the right environment, heart muscle cells. This type of stem cell is commonly referred to as an "adult stem cell," to distinguish it from one derived from an embryo.

The human embryo, formed by union of sperm and egg, begins as a single cell. This cell is phenomenally flexible, pos-

245

sessed of potential to turn into a liver cell, a brain cell, a muscle cell, and every other kind of complex tissue that makes up the 100 trillion cells of the adult human being. The weight of current evidence is that the potential of an embryonic stem cell for sustained replication and the ability to become virtually any cell type exceeds that of an adult stem cell. By definition, however, a human embryonic stem cell can be derived only from an early embryo—not necessarily at the single-cell stage, but while the embryo is still only a small compact ball of cells smaller than the dot on this letter *i.*

But Dolly was derived neither from an embryonic stem cell nor an adult stem cell. The truly dramatic and unexpected aspect of Dolly's creation is that she came about by a method wholly unprecedented in mammals, one that does not occur in nature. As shown in Figure A.1, this process, technically known as somatic cell nuclear transfer (SCNT), began with a single cell derived from the udder of a mature sheep (the donor). The nucleus of that cell, carrying the complete DNA of that donor sheep, was then removed and inserted into the rich environment of proteins and signaling molecules found in the cytoplasm of an egg cell.

That egg cell had previously had its nucleus completely removed, so it could not provide the needed genetic instructions, only the environment for those instructions to be recognized and carried out. Placed into that primordial embrace, the DNA from the udder cell effectively went back through time, erasing all of the specific changes its DNA packaging had experienced on the way to becoming a very specialized cell involved in milk production. The udder cell nucleus reverted back to its primitive

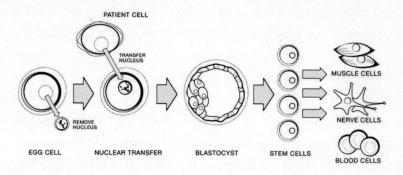

Figure A.1 The process of somatic cell nuclear transfer (SCNT).

undifferentiated state. Implanting that cell back into the womb of a sheep then gave rise to Dolly, whose nuclear DNA was identical to that of the original donor sheep.

The scientific and medical research world was electrified by the utterly unexpected flexibility of the genome instruction book. Building on that revelation, scientists now see the study of stem cells as a real opportunity to learn how a single cell can become a liver cell, a kidney cell, or a brain cell. Of course, many of these basic questions are being answered by studying stem cells from animals, where the ethical concerns are much more limited. The real excitement about medical benefits from stem cell research, however, is the potential, albeit still un-proven, to use this approach to develop new therapies. Many chronic diseases arise because a certain cell type dies prema-turely. If your daughter has juvenile onset (type I) diabetes, it is because the cells in her pancreas that normally secrete insulin have undergone an immune attack by the body and have died off. If your father has Parkinson's disease, it is because neurons in a particular part of his brain, the substantia nigra, have died prematurely, resulting in a disruption of the normal circuits that control motor function. If your cousin is on the transplant list for a liver, or a kidney, or a heart, it is because those organs have sustained severe enough damage that they cannot repair themselves.

If a means could be found by which we could regenerate these damaged tissues or organs, then many currently progressive and fatal chronic illnesses could be effectively treated, or even cured. For that reason, "regenerative medicine" is a topic of enormous interest in medical research. At the present time, the study of stem cells seems to offer the greatest promise to realize this dream.

A furious social, ethical, and political debate has arisen, however, surrounding the study of human stem cells. The intensity of emotion, the passion of the various perspectives, and the clash of views are almost unprecedented, and often the scientific details have been lost in the storm.

First of all, few would argue that the therapeutic use of adult stem cells presents any major new ethical dilemmas. Such cells can be derived from the tissues of an individual who is already living. The desired scenario would then be to convince that cell to morph into the type of cell needed to treat that person's illness. If we knew how, for instance, to convert a few bone marrow stem cells into a very large number of liver cells, then a liver "autotransplant" might be accomplished simply by utilizing a patient's own marrow.

While there have been some encouraging steps in that direction, and a very significant investment is being made in pursuing adult stem cell research, at the present time we lack confirmatory evidence that the repertory of adult stem cells that humans possess will be sufficient to meet many of the needs of people with chronic illnesses. Human embryonic stem cells, or alternatively the use of somatic cell nuclear transfer, are therefore being seriously explored as potential alternatives.

Stem cells derived from a human embryo should have the ultimate potential to form any type of tissue (after all, they do so naturally in the course of events). But here is where profound ethical questions are being raised, and rightly so. An embryo formed by the union of a human sperm and egg is a potential human life. Deriving stem cells from an embryo generally results in the destruction of the embryo (though a few methods have been proposed that might still allow its survival). If one believes unequivocally that life begins at conception, and that human life is sacred from that very moment onward, then this would be an unacceptable form of research or medical care.

Reasonable people will disagree, often passionately, about the appropriateness of such research. Where one comes down on the spectrum of acceptable versus unacceptable is strongly influenced by one's answers to the following questions.

Does human life begin at conception?
Scientists, philosophers, and theologians have debated for centuries the point at which life actually begins. Deriving more information about the actual anatomical and molecular steps involved in the early development of the human embryo has not really helped with those debates, as this is not really a

scientific question. For centuries, different definitions of the beginning of life have been offered by different cultures and faith traditions, and even today different faiths use different milestones to mark the entrance of the soul into the human fetus.

From a biologist's perspective, the steps that follow the union of sperm and egg occur in a highly predictable order, leading to increasing complexity, and with no sharp boundaries between phases. There is therefore no convenient biological dividing line between a human being and an embryonic form that might be called "not quite there yet." Some have argued that truly human existence cannot exist without a nervous system, so the fetal development of the "primitive streak" (the earliest anatomic precursor of the spinal cord, which generally appears at about day fifteen) could potentially be used as such a marker. Others argue that the potentiality of the embryo to develop a nervous system exists from the moment of conception, and it is not relevant whether or not that potentiality has actually been realized in the formation of any particular anatomic structure.

Interesting light has been shed on this issue from the existence of identical twins, who develop from a single fertilized egg. Very early in development (presumably at the two-cell stage), the embryo comes apart, resulting in two distinct embryos with identical DNA sequences. No theologian would argue that identical twins lack souls, or that they share a single soul. In these cases, therefore, the insistence that the spiritual nature of a person is uniquely defined at the very moment of conception encounters a difficulty.

Are there any circumstances under which it would be justifiable to derive stem cells from human embryos?

Those who feel strongly that human life begins at conception, and that from that very moment the embryo deserves the full moral status of an adult human being, will generally answer no to this question. Their stance would be ethically consistent. However, it must be pointed out that many such individuals have chosen to look the other way, or at least adopt a position of moral relativism, in another circumstance where human embryos are being destroyed.

That is the process of in vitro fertilization (IVF), now widely available for infertile couples, and widely embraced as a solution to a terrible heartache. In this procedure, eggs are harvested from the mother after a hormonal treatment that results in many eggs being released at once. The eggs are fertilized in a petri dish with the prospective father's sperm. Embryos are observed for three to six days to assess whether they are developing normally, and then a small number (usually one or two) are implanted into the mother, in hopes of achieving a pregnancy.

In most instances, there are more embryos available than can safely be implanted. The spare embryos are often frozen. In the United States alone there are hundreds of thousands of such frozen embryos currently stored in freezers, and that number continues to grow. While actual adoption of these embryos by other couples has resulted in a small number of them giving rise to pregnancies, there is no question that the vast majority of these embryos will ultimately be discarded. A strict stance in opposition to the destruction of human embryos

251

under any circumstance would appear, therefore, to require opposition to in vitro fertilization. A demand that all embryos generated by IVF be implanted has also been proposed, but that would increase the risk of fetal death from multiple pregnancy. There is really no easy way around this dilemma at the present time.

Many observers who are otherwise opposed to human embryo research have argued, however, that despite the likely ultimate destruction of excess embryos after IVF, the desire of a couple to have a child is such a strong moral good that it justifies the procedure. That may well be a defensible position, but if so, it challenges the principle that the inevitable destruction of human embryos should be avoided at all costs, no matter what the potential benefits.

This circumstance raises the question being asked by many: if procedures could be set up to ensure that no in vitro fertilization procedure was ever undertaken with the explicit intent of generating embryos for research, and if medical research were then restricted only to those embryos that were left over after IVF and clearly destined for destruction, would that be a moral violation?

SOMATIC CELL NUCLEAR TRANSFER IS FUNDAMENTALLY DIFFERENT

The good news is that these furious debates about stem cells cultivated from human embryos may ultimately turn out to be unnecessary, as another, less ethically challenging pathway may provide even more powerful medical breakthroughs. I refer to the same process of somatic cell nuclear transfer (SCNT) that resulted in Dolly the sheep.

It is intensely regrettable that the product of SCNT has been equated both in terminology and in moral argument with the generation of stem cells from a human embryo derived from the union of sperm and egg. This equivalency, arrived at very early in the public debate and now adhered to almost slavishly by most participants, ignores the profound difference between the way in which these two entities are generated. The SCNT procedure has potentially a much greater likelihood of providing medical benefit, and so it is particularly important that we try to untangle the confusion that has surrounded this process.

As described above and shown in Figure A.1, SCNT involves no sperm-and-egg fusion. Instead, the DNA instruction book is derived from a single cell from the skin or some other tissue of a living animal. (For Dolly, it happened to be the udder, but it could be almost anything.) Virtually everyone would agree that an initial donor skin cell has no particular moral value; after all, we shed millions of them every day. Similarly, the enucleated egg cell, having lost all of its own DNA, has *no* potential of ever becoming a living organism, and therefore also does not seem deserving of moral status. Putting these two entities together creates a cell that does not occur naturally but has great ultimate potential. But should we call it a human being?

If one argues that the sheer fact of its ultimate potentiality deserves that claim, then why would not that same argument apply to the skin cell before it had been manipulated? It had potential too.

Over the course of the next few years it is likely that scientists will discover the signals that are contained within the egg cell cytoplasm that allow the skin cell nucleus to erase its history and recover its remarkable potential to turn into many different types of tissues. Thus, it is likely that within a few years this process will no longer require the egg at all, but will be achieved by dropping any type of cell from an individual donor into an appropriate cocktail of signaling molecules. At what point, along that long series of steps, should the moral status of a human being be assigned? Wouldn't the result of this procedure resemble an adult stem cell more than an embryonic stem cell?

The fuss about SCNT derives from the fact that this bizarre fusion of udder cell and enucleated egg cell ultimately resulted in Dolly. That happened only because the product of SCNT was intentionally placed back into a sheep uterus, hardly something that could happen by accident. Similar steps have now been taken for many other mammals, including cows, horses, cats, and dogs. This so-called reproductive cloning may have even been attempted in humans by a couple of fringe research groups, one of which (the Raelians) is led by an individual who wears silver jumpsuits and claims to have been abducted by aliens (not exactly credentials for a scientist). Scientists, ethicists, theologians, and lawmakers are essentially unanimous that reproductive cloning of a human being should not be undertaken under any circumstances. While a major reason for this stance is based on strong moral and theological objections to making human copies in this unnatural way, other major objections are based upon safety considerations, since reproductive cloning of every other mammal has been shown to be an incredibly inefficient and disaster-prone effort, with most clones resulting in miscarriage or early infant death. The few clones that have survived beyond birth have been almost uniformly abnormal in some way, including Dolly herself (she suffered from arthritis and obesity).

Given those conclusions, it would be entirely appropriate to demand that the product of human somatic cell nuclear transfer never be reimplanted into the womb of a host mother. Virtually everyone can agree about that. The battle revolves around whether human SCNT should be undertaken under any other circumstances, when there is no intent whatsoever to produce an intact human being. The stakes are potentially very high. If you are dying of Parkinson's disease, it is not stem cells from some other donor that you need, it is your own. After all, over many decades we have learned through the science of organ transplantation that putting cells from another individual into a recipient predictably creates a destructive rejection response, which can usually be minimized only by careful tissue matching between donor and recipient, and the posttransplant use of powerful immunosuppressive drugs, with all of the complications these entail. Many of the scenarios that advocate the use of anonymous embryonic stem cells from unrelated donors to treat various diseases fly in the face of this long experience.

It would be far better, therefore, if stem cells were genetically identical to their recipient. This is, of course, exactly the outcome that would occur after SCNT. (This is also referred to as "therapeutic cloning," although that term carries enough rhetorical baggage to render it now almost useless.) It is hard for an objective observer to argue that this will not be, in the long run, a promising pathway toward treatment for a long list of debilitating and ultimately fatal diseases. It behooves us to look very carefully, therefore, at the moral objections to such a potentially beneficial process and assess whether they deserve the weight they are being given in some quarters.

I would argue that the immediate product of a skin cell and an enucleated egg cell fall short of the moral status of the union of sperm and egg. The former is a creation in the laboratory that does not occur in nature, and is not part of God's plan to create a human individual. The latter is very much God's plan, carried out through the millennia by our own species and many others.

Like virtually everyone else, I am strongly opposed to the idea of human reproductive cloning. Implanting the product of human SCNT into a uterus is profoundly immoral and ought to be opposed on the strongest possible grounds. On the other hand, protocols are already being developed to convince a single cell that has been derived from SCNT to be converted into a cell that senses glucose levels and secretes insulin, without going through any of the other steps of embryonic and fetal development. If such steps can result in tissue-matched cells that cure juvenile diabetes, why would that not be a morally acceptable procedure?

There is no question that the science in this field will continue to move rapidly. While the ultimate medical benefits of stem cell research remain undefined, there is great potential in them. Opposing all research of this kind means the ethical mandate to alleviate suffering has been trumped absolutely by other perceived moral obligations. For some believers, that may be a defensible stance, but it should be arrived at only after a complete consideration of the facts. Anyone who portrays this issue as a simple battle between belief and atheism does a disservice to the complexity of the issues.

BEYOND MEDICINE

My morning newspaper recently included an analysis of various challenges facing the president of the United States. This particular story, coming at a time when things weren't going very well for the Commander-in-Chief, included a quote from someone identified as a political consultant and friend: "I've never seen the president burdened by the presidency. He's built to deal with really big events. It's in his DNA."

While the president's friend may have intended his comment as a contemporary witticism, it's entirely possible that he meant it.

What is the real evidence for heritability of human behaviors and personality traits? And will the genomics revolution lead us into new ethical questions because of it? How does one really assess the roles of heredity and environment in such complex human characteristics? Many erudite treatises have been written on this subject. But long before Darwin, Mendel, Watson, Crick, and all the rest, observant humans had already figured out that nature has provided us with a wonderful opportunity to assess the role of inheritance in many different aspects of human existence. That opportunity is provided by identical twins.

If you have encountered a pair of identical twins, you will

Personality Trait	Heritability Estimate
General cognitive ability	50%
Extroversion	54%
Agreeableness	42%
Conscientiousness	49%
Neuroticism	48%
Openness	57%
Aggression	38%
Traditionalism	54%

Table A.1 Estimates of the Percentage of Various Human Personality Traits that can be Ascribed to Heredity, from T. J. Bouchard and M. McGue, "Genetic and Environmental Influences on Human Psychological Differences," *J. Neurobiol.* 54 (2003):4–45. Each of the traits listed here has a strict definition in the science of personality analysis.

agree that they share remarkable physical resemblance, as well as other traits such as pitch of voice and even certain mannerisms. However, if you get to know them well, you will find that they have distinct personalities. Scientists have studied identical twins for centuries in order to assess the contributions of nature and nurture to a wide variety of human characteristics.

An even more unbiased careful analysis can be done on identical twins who were adopted to different homes at birth, and therefore had totally different childhood environments. Such twin studies allow an estimate of the heritability of any particular trait without in any way determining its actual molecular basis. Table A.1 shows some examples of the estimates of the proportion of a particular trait contributed by heredity, based on twin studies. For various methodological reasons, however, these should not be taken as precise.

These studies lead to the conclusion that heredity is important in many of these personality traits. That will not surprise any of us who live within families. We should therefore not be too shaken up by the fact that certain molecular details about the mechanism of heritability are beginning to be unearthed through the study of the genome. But we are.

It is one thing to say you have your grandmother's eyes or your grandfather's temper. It is another to say that those things came about because you have a certain T or C in a particular place in your genome, which you may or may not have passed on to your children. Though genetic research on human behavior holds the exciting promise of improved interventions in psychiatric illness, this research is also somehow upsetting, as it seems to tread dangerously close to threatening our free will, our individuality, and maybe even our spirituality.

We need to get used to this, however. The molecular definition of certain human behaviors is already happening. Several groups have published papers in the scientific literature indicating that common variants in a receptor for the neurotransmitter dopamine are associated with an individual's score on the "novelty seeking" trait in a standardized personality test. This receptor variant, however, contributes only a very small proportion of the variability in this particular trait. While the result may be statistically interesting, it is essentially irrelevant for the individual.

Other groups have identified a variant in a transporter for

another neurotransmitter, serotonin, that is associated with anxiety. That same transporter variant has also been reported to correlate statistically with whether or not an individual experiences significant depression after a major life stress event. If correct, this would be an example of a gene-environment interaction.

An area of particularly strong public interest is the genetic basis of homosexuality. Evidence from twin studies does in fact support the conclusion that heritable factors play a role in male homosexuality. However, the likelihood that the identical twin of a homosexual male will also be gay is about 20 percent (compared with 2–4 percent of males in the general population), indicating that sexual orientation is genetically influenced but not hardwired by DNA, and that whatever genes are involved represent predispositions, not predeterminations.

Of the many aspects of human individuality that are most likely to cause controversy, none could be more explosive than intelligence. While disagreement about how to define intelligence and how to measure it remain a hot topic in social science, and while the various available IQ tests clearly measure a bit of learning and culture, not just general cognitive ability, there is clearly a strong heritable component in this human attribute (Table A.1). At this writing, no specific DNA variant has yet been shown to play a role in IQ. It is likely, however, that there will eventually be dozens of such variants, once our methods are good enough to discover them. As with other aspects of human behavior, no single variant is likely to make more than a tiny contribution (perhaps one to two IQ points).

Could criminality even be influenced by inherited susceptibilities? In a way that is both obvious to everyone but not usually considered in quite this context, we already know this to be true. Half of our population carries a specific genetic variant that makes them sixteen times more likely to end up in jail than the other half. I am, of course, referring to the Y chromosome carried by males. The knowledge of that association, however, has not undermined our social fabric, nor has it been used successfully as a criminal defense by guilty males.

But putting that obvious point aside, it is indeed possible that other modest contributions to antisocial behavior will be identified in the genome. A particularly interesting example has already appeared, beginning with the observation of a single family in the Netherlands where the incidence of antisocial and criminal behavior among many of the males in the family stood out dramatically, and was consistent with the pattern of inheritance one might see for a gene on the X chromosome.

Careful study of this Dutch family revealed that there was an inactivating mutation in the gene for monoamine oxidase A (MAOA) on the X chromosome, and all of the males who had exhibited antisocial behavior carried the mutation. This could simply be a rare event with no broader significance, but it turns out that the normal MAOA gene has two different versions, a high expresser and a low expresser. While there is no overall evidence that low-expresser males have a higher frequency of interactions with the law, a careful study in Australia that looked at boys who were abused as children concluded that those who carried the low-expresser MAOA had a substantially higher frequency of antisocial and criminal behavior as adults. Here again may be an example of gene-environment interaction: the genetic susceptibility conferred by MAOA becomes apparent only when the environmental experience of child abuse is added to the picture. But even in this situation, the findings were significant only on a statistical basis. There were plenty of individual exceptions to the rule.

A few years ago, I saw an article in a religious periodical asking the question whether individual spirituality might even be genetic. I smiled, thinking that now I had heard the ultimate in genetic determinism. But perhaps I was too hasty; it is not impossible to imagine that certain personality types, themselves based upon weakly inherited factors, may be more prone to accept the possibility of God than others. A recent twin study suggested just that, though as usual one must add the caveat that the observed effect of heredity was quite weak.

The question of the genetics of spirituality has recently achieved wide attention with the publication of a book called *The God Gene*,[3] by the same researcher who has also published findings on novelty seeking, anxiety, and male homosexuality. The book grabbed headlines, and even the cover of *Time* magazine, but a careful reading indicated that the title was wildly overstated.

The researcher utilized personality testing to deduce that a trait called "self-transcendence" showed heritability in families and twins. This characteristic was associated with an individual's ability to accept things that cannot be directly proven or measured. The demonstration that such a personality parameter might have heritable characteristics is, in itself, not surprising, since most personality traits do seem to have such properties. But the investigator went on to claim that a variant in a particular gene, VMAT2, was associated with a higher score on the self-transcendence scale. As none of his data has been peer reviewed or published in the scientific literature, most experts have greeted the book with considerable skepticism.

A reviewer in *Scientific American* quipped that the appropriate title for the book should have been *A Gene That Accounts for Less Than One Percent of the Variance Found in Scores on Psychological Questionnaires Designed to Measure a Factor Called Self-Transcendence, Which Can Signify Everything from Belonging to the Green Party to Believing in ESP, According to One Unpublished, Unreplicated Study.*

To summarize this section: There is an inescapable component of heritability to many human behavioral traits. For virtually none of them is heredity ever close to predictive. Environment, particularly childhood experiences, and the prominent role of individual free will choices have a profound effect on us. Scientists will discover an increasing level of molecular detail about the inherited factors that undergird our personalities, but that should not lead us to overestimate their quantitative contribution. Yes, we have all been dealt a particular set of cards, and the cards will eventually be revealed. But how we play the hand is up to us.

ENHANCEMENT

The science fiction movie *GATTACA* depicts a future society in which the genetic factors for susceptibility to disease and human behavior traits have all been identified, and are used diagnostically to optimize the outcome of a mating. In this chilling future vision, society has abandoned all of its individual freedoms and allowed individuals to be channeled into particular occupations and life experiences based upon the DNA they carry. The movie's premise, that genetic determinism can be so accurate that a society would tolerate this kind of circumstance, is undercut by the fact that its hero (born outside the system) still manages to outperform all of the enhanced individuals, who smoke, drink, and murder one another.

Does this sort of science fiction deserve any credibility? Certainly the topic of future human enhancement is taken seriously by many, including some prominent scientists. I sat in the audience in 2000 at a "Millennium Evening" at the White House, attended by the president, when no less a scientific eminence than Stephen Hawking advocated that it was time for humanity to take charge of evolution, and to plan a program of systematic self-improvement of the species. While in a certain way one can understand Hawking's motivation, afflicted as he is with a debilitating neurological disease, I found his proposal chilling. Who decides what is an "improvement"? How disastrous might it be to reengineer our species, only to discover we had lost something critical (like resistance to an emerging disease) along the way? And how would such wholesale redesign affect our relationship with our Creator?

The good news is that such scenarios are a very long way off, if indeed they are ever to become possible. But there are other aspects of human enhancement that are closer at hand, and more appropriate for consideration here.

First, let's admit that enhancement is not an easy concept to define precisely. Nor is there a bright line between treating illness and enhancing function. Consider obesity, for instance. Morbid obesity is certainly associated with a host of serious medical problems, and is an appropriate topic for medical research and prevention and treatment. On the other hand, developing a means to allow persons of normal weight to achieve ultraslender supermodel status could hardly be called a medical triumph. Yet the body-weight spectrum between these two extremes is a continuous one, and there is no easy way to determine when you have crossed the line.

Before jumping to the conclusion that enhancement of ourselves or our children is unacceptable and dangerous territory, it is well to remember that in many instances we are already doing it, and even insisting on it. We are considered irresponsible parents if we do not assure that our children obtain appropriate immunizations against infectious diseases. Make no mistake: an immunization is most definitely an enhancement, as it leads to the proliferation of certain clones of immune cells, and even rearrangements of DNA.

Similarly, fluoridated water, music lessons, and orthodontics are generally considered desirable enhancements. Regular exercise, an enhancement of our physical status, is a laudable activity. And while coloring one's hair or taking advantage of cosmetic surgery may be considered vain, most would not call such actions immoral.

On the other hand, certain currently available enhancements are considered to have questionable moral status, though part of the judgment depends on the context. The use of injectable growth hormone is acceptable for kids with a pituitary deficiency, but most would view it as inappropriate for parents who simply wish to increase the natural height of their children. Similarly, while the use of the blood-enhancing hormone erythropoietin has been a godsend for individuals with kidney failure, its use by athletes is considered both immoral and illegal. For another example relating to athletics, the use of the growth factor IGF-1 shows great promise in animal studies to increase muscle mass, and would be very difficult to detect by current monitoring systems. Most would consider this just as unacceptable as steroids in the athletic setting. But IGF-1 appears potentially able also to slow down the aging process. If that turns out to be true, would this use also be immoral?

None of the examples cited so far have actually altered the "germ-line" DNA (the DNA that is passed from parent to child) of the individual, and it is highly unlikely that such experiments on humans will be undertaken anytime in the near future. While this is routinely done in animal experiments, there are serious safety issues that would preclude its application to humans, given that the negative consequences of such a manipulation might not be apparent for several generations to come. Clearly the future offspring whose genomes were manipulated would not have had the opportunity to give consent. From an ethical perspective, therefore, germ-line manipulations of human beings are likely to remain off the table for a very long time. The one possible exception to that might be if one could construct a truly artificial human chromosome to carry extra material, but equip that chromosome with a self-destruct mechanism if something started to go wrong. However, we are still a very long way away from implementing any such protocol, even in animals.

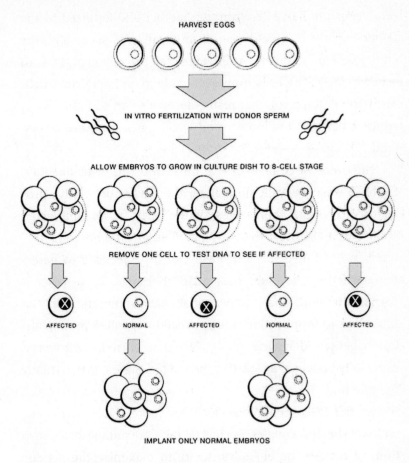

Figure A.2 Preimplantation genetic diagnosis (PGD).

Does this mean, then, that any fears about manipulation of the human gene pool are overblown? Yes, if you're talking about genetic engineering of the germ line to create new DNA structures. But no, if you're talking about the *GATTACA* scenario of embryo selection. This high-tech but increasingly widespread practice has added a new twist to in vitro fertilization. As shown in Figure A.2, at the time of in vitro fertilization a dozen or so eggs are harvested from the mother and fertilized by the father's sperm in a petri dish. If fertilization is successful, the embryos begin to divide. At the eight-cell stage, it is possible to remove one of the cells from each embryo and perform a DNA test upon it. Based on that result, decisions can be made about which embryos to reimplant and which ones to freeze or discard.

Hundreds of couples at risk for serious diseases like Tay-Sachs disease or cystic fibrosis have already utilized this procedure to assure themselves of the birth of an unaffected child. But a DNA test that reveals whether an embryo is destined to have Tay-Sachs disease can also be used to determine whether the embryo is male or female, or whether it carries an adult-onset-disease risk like a mutation in the BRCA1 gene. Application of this procedure, called preimplantation genetic diagnosis (PGD), has thus stirred controversy, especially because, at least in the United States, it is virtually unregulated.

As the PGD technology becomes more widely available, will well-heeled couples decide to take advantage of it, in a form of homemade eugenics, to try to maximize the genetic endowment of their offspring, in order to try to achieve the optimum mix of the parents' genomes? Will they try to weed out less desirable variants and make sure certain traits are passed along?

There is a statistical problem with this approach. The kinds of attributes that parents might want to enhance are generally controlled by multiple genes. Yet getting both Mom's best version and Dad's best version for any given gene will happen

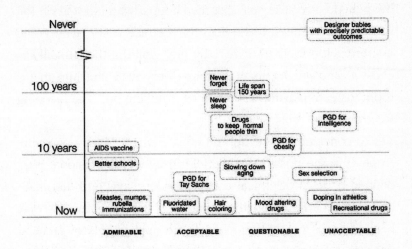

Figure A.3 A graphical depiction of various enhancement scenarios. While not all would agree on the precise likelihood of occurrence or degree of ethical concern for each example, this diagram may help to prioritize situations in the lower right quadrant as being of most immediate importance.

only in one out of four embryos. If two genes are to be optimized, it will take sixteen embryos (on average) to find one that meets that requirement. To optimize for ten genes, it would take more than a million embryos! Since that is substantially more than the total number of eggs a woman can produce in her lifetime, the silliness of the scenario becomes immediately apparent.

There is another good reason why the scenario is silly, however. Even for that one-in-a-million embryo, the choice of ten genes for intelligence, musical ability, or athletic prowess would be likely to skew the odds only by a small amount. Furthermore, none of these genes would operate in isolation. The critical importance of childhood upbringing, education, and discipline would not be obviated by a slightly optimized throw of

269

the genetic dice. The self-absorbed couple who insisted on the use of such genetic technology to produce a son who could quarterback a football team, play first violin in the student orchestra, and get A+ in math might very well find him in his room instead, playing video games, smoking pot, and listening to heavy metal music.

To conclude this section on enhancement, it may be useful to place some possible scenarios on a two-dimensional plot, defined by level of ethical concern on one axis and the likelihood of occurrence on the other. That plot (Figure A.3) may help us focus our attention on those applications of greatest concern, which fall in the lower right quadrant.

CONCLUSION

This survey of some of the ethical dilemmas associated with coming advances in genomics and related fields is by no means exhaustive. New dilemmas seem to be born every day, and some of the ones described in this Appendix may fade away. For those issues that represent real ethical challenges, and not artificial and unrealistic scenarios, how are we as a society to arrive at conclusions?

First of all, it would be a mistake to simply leave those decisions to the scientists. Scientists have a critical role to play in such debates, since they possess special expertise that may enable a clear distinction of what is possible and what is not. But scientists can't be the only ones at the table. Scientists by their nature are hungry to explore the unknown. Their moral sense is in general no more or less well developed than that of other groups, and they are unavoidably afflicted by a potential conflict of interest that may cause them to resent boundaries set by nonscientists. Therefore, a wide variety of other perspectives must be represented at the table. The burden is heavy upon those participating in such debates, however, to educate themselves about the scientific facts. As the current debate about stem cells has taught us, hardened positions can sometimes develop long before the nuances of the science have become clear, to the detriment of the potential for real dialogue.

Does a person's grounding in one of the great world faiths assist his or her ability to resolve these moral and ethical dilemmas? Professional bioethicists would generally say no, since as we have already noted, the principles of ethics such as autonomy, beneficence, nonmaleficence, and justice are held true by believers and nonbelievers alike. On the other hand, given the uncertain ethical grounding of the postmodernist era, which discounts the existence of absolute truth, ethics grounded on specific principles of faith can provide a certain foundational strength that may otherwise be lacking. I hesitate, however, to advocate very strongly for faith-based bioethics. The obvious danger is the historical record that believers can and will sometimes utilize their faith in a way never intended by God, and to move from loving concern to self-righteousness, demagoguery, and extremism.

No doubt those who conducted the Inquisition thought themselves to be carrying out a highly ethical activity, as did those who executed witches in Salem, Massachusetts. In our time, Islamic suicide bombers and assassins of abortion-clinic doctors no doubt are also convinced of their moral righteousness. As we face challenging dilemmas wrought by science in the future, let us bring every right and noble tradition of the world, tried and proven true through the centuries, to the table. But let us not imagine that every individual interpretation of those great truths will be honorable.

Is the science of genetics and genomics beginning to allow us to "play God"? That phrase is the one most commonly used by those expressing concern about these advances, even when the speaker is a nonbeliever. Clearly the concern would be lessened if we could count on human beings to play God as God does, with infinite love and benevolence. Our track record is not so good. Difficult decisions arise when a conflict appears between the mandate to heal and the moral obligation to do no harm. But we have no alternative but to face those dilemmas head-on, attempt to understand all of the nuances, include the perspectives of all the stakeholders, and try to reach a consensus. The need to succeed at these endeavors is just one more compelling reason why the current battles between the scientific and spiritual worldviews need to be resolved—we desperately need both voices to be at the table, and not to be shouting at each other.

NOTES

Introduction

1. R. Dawkins, "Is Science a Religion?" *The Humanist* 57 (1997): 26–29.

2. H. R. Morris, *The Long War Against God* (New York: Master Books, 2000).

Chapter 1: From Atheism to Belief

1. C. S. Lewis, "The Poison of Subjectivism," in *C. S. Lewis, Christian Reflections,* edited by Walter Hooper (Grand Rapids: Eerdmans, 1967), 77.

2. J. Chittister in F. Franck, J. Roze, and R. Connolly (eds.), *What Does It Mean To Be Human? Reverence for Life Reaffirmed by Responses from Around the World* (New York: St. Martin's Griffin, 2000), 151.

3.   C. S. Lewis, *Mere Christianity* (Westwood: Barbour and Company, 1952), 21.

4.   S. Vanauken, *A Severe Mercy* (New York: HarperCollins, 1980), 100.

Chapter 2: The War of the Worldviews

1.   P. Tillich, *The Dynamics of Faith* (New York: Harper & Row, 1957), 20.

2.   C. S. Lewis, *Surprised by Joy* (New York: Harcourt Brace, 1955), 17.

3.   S. Freud, *Totem and Taboo* (New York: W. W. Norton, 1962).

4.   A. Nicholi, *The Question of God* (New York: The Free Press, 2002).

5.   C. S. Lewis, *Mere Christianity* (Westwood: Barbour and Company, 1952), 115.

6.   A. Dillard, *Teaching a Stone to Talk* (New York: Harper-Perennial, 1992), 87–89.

7.   Voltaire quoted in Alister McGrath, *The Twilight of Atheism* (New York: Doubleday, 2004), 26.

8.   C. S. Lewis, *The Problem of Pain* (New York: MacMillan, 1962), 23.

9.   Ibid., 25.

10.   Ibid., 35.

11.   Ibid., 83.

12.   D. Bonhoeffer, *Letters and Papers from Prison* (New York: Touchstone, 1997), 47.

13.   C. S. Lewis, *Miracles: A Preliminary Study* (New York: MacMillan, 1960), 3.

14. Ibid., 167.

15. J. Polkinghorne, *Science and Theology—An Introduction* (Minneapolis: Fortress Press, 1998), 93.

Chapter 3: The Origins of the Universe

1. E. Wigner, "The Unreasonable Effectiveness of Mathematics in the Natural Sciences," *Communications on Pure and Applied Mathematics* 13, no. 1 (Feb. 1960).

2. S. Hawking, *A Brief History of Time* (New York: Bantam Press, 1998), 210.

3. R. Jastrow, *God and the Astronomers* (New York: W. W. Norton, 1992), 107.

4. Ibid., 14.

5. Hawking, *Brief History,* 138.

6. For a thorough and rigorously mathematical enumeration of these arguments, see J. D. Barrow and F. J. Tipler, *The Anthropic Cosmological Principle* (New York: Oxford University Press, 1986).

7. I. G. Barbour, *When Science Meets Religion* (New York: HarperCollins, 2000).

8. Hawking, *Brief History,* 144.

9. F. Dyson cited in Barrow and Tipler, *Principle,* 318.

10. A. Penzias quoted in M. Browne, "Clues to the Universe's Origin Expected," *New York Times,* March 12, 1978.

11. J. Leslie, *Universes* (New York: Routledge, 1989).

12. Hawking, *Brief History,* 63.

13. Saint Augustine, *The Literal Meaning of Genesis,* translated and annotated by John Hammond Taylor, S.J. (New York: Newman Press, 1982), 1:41.

Chapter 4: Life on Earth

1. W. Paley, *The Works of William Paley,* edited by Victor Nuovo and Carol Keene (New York: Thoemmes Continuum, 1998).

2. C. R. Woese, "A New Biology for A New Century," *Microbiology and Molecular Biology Reviews* 68 (2004): 173–86.

3. D. Falk, *Coming to Peace with Science* (Downers Grove: Intervarsity Press, 2004).

4. C. R. Darwin, *The Origin of Species* (New York: Penguin, 1958), 456.

5. B. B. Warfield, "On the Antiquity and the Unity of the Human Race," *Princeton Theological Review* 9 (1911): 1–25.

6. Darwin, *Origin,* 452.

7. Ibid., 459.

8. C. R. Darwin, quoted in Kenneth R. Miller, *Finding Darwin's God* (New York: HarperCollins, 1999), 287.

Chapter 5: Deciphering God's Instruction Book

1. R. Cook-Deegan, *The Gene Wars* (New York: Norton, 1994).

2. J. E. Bishop and M. Waldholz, *Genome* (New York: Simon & Schuster, 1990); K. Davies, *Cracking the Genome* (New York: Free Press, 2001); J. Sulston and G. Ferry, *The Common Thread* (Washington: Joseph Henry Press, 2002); I. Wickelgren, *The Gene Masters* (New York: Times Books, 2002); J. Shreeve, *The Genome War* (New York: Knopf, 2004).

3. T. Dobzhansky, "Nothing in Biology Makes Sense Except in the Light of Evolution," *American Biology Teacher* 35 (1973): 125–29.

Chapter 6: Genesis, Galileo, and Darwin

1. Saint Augustine, *The City of God* XI.6.
2. Saint Augustine, *The Literal Meaning of Genesis* 20:40.
3. A. D. White, *A History of the Warfare of Science with Theology in Christendom* (New York, 1898); see www.santafe.edu/~shalizi/White.
4. See http://en.wikipedia.org/wiki/Galileo_Galilei.
5. Augustine, *Genesis* 19:39.
6. Galileo, letter to Grand Duchess Christina, 1615.

Chapter 7: Option 1: Atheism and Agnosticism

1. Saint Augustine, *Confessions* I.i.1.
2. E. O. Wilson, *On Human Nature* (Cambridge: Harvard University Press, 1978), 192.
3. R. Dawkins, "Is Science a Religion?" *The Humanist* 57 (1997): 26–29.
4. S. Clemens, *Following the Equator* (1897).
5. R. Dawkins, *The Selfish Gene,* 2nd ed. (Oxford: Oxford University Press, 1989), 198.
6. Ibid., 200–201.
7. S. J. Gould, "Impeaching a Self-Appointed Judge" (review of Phillip Johnson's *Darwin on Trial*), *Scientific American* 267 (1992):118–21.
8. T. H. Huxley, quoted in *The Encyclopedia of Religion and Ethics,* edited by James Hastings (1908).
9. See http://en.wikipedia.org/wiki/Charles_Darwin's_views_on_religion.

Chapter 8: Option 2: Creationism

1. B. B. Warfield, *Selected Shorter Writings* (Phillipsburg: PRR Publishing, 1970), 463–65.

Chapter 9: Option 3: Intelligent Design

1. For additional details of these arguments, see W. A. Dembski and M. Ruse, eds., *Debating Design: From Darwin to DNA* (Cambridge: Cambridge University Press, 2004).

2. This example is covered in much greater detail in K. R. Miller, *Finding Darwin's God* (New York: HarperCollins, 1999), 152–61.

3. C. Darwin, *The Origin of Species* (New York: Penguin, 1958), 171.

4. K. R. Miller, "The Flagellum Unspun," in Dembski and Ruse, *Debating Design,* 81–97.

5. Darwin, *Origin,* 175.

6. W. A. Dembski, "Becoming a Disciplined Science: Prospects, Pitfalls, and Reality Check for ID" (keynote address, Research and Progress in Intelligent Design Conference, Biola University, La Mirada, Calif., Oct. 25, 2002).

7. W. A. Dembski, *The Design Revolution* (Downers Grove: Intervarsity, 2004), 282.

8. R. Dawkins, *River Out of Eden: A Darwinian View of Life* (London: Weidenfeld and Nicholson, 1995).

Chapter 10: Option 4: BioLogos

1. See, for example, R. C. Newman, "Some Problems for Theistic Evolution," *Perspectives on Science and Christian Faith* 55 (2003): 117–28.

2.   Pope John Paul II, "Message to the Pontifical Academy of Sciences: On Evolution," Oct. 22, 1996.

3.   Cardinal Christoph Schönborn, "Finding Design in Nature," *New York Times,* July 7, 2005.

4.   T. Dobzhansky, "Nothing in Biology Makes Sense Except in the Light of Evolution," *American Biology Teacher* 35 (1973): 125–29.

5.   C. S. Lewis, *The Problem of Pain* (New York: Simon & Schuster, 1996), 68–71.

Chapter 11: Truth Seekers

1.   C. S. Lewis, *Mere Christianity* (Westwood: Barbour and Company, 1952), 50.

2.   L. Strobel, *The Case for Christ* (Grand Rapids: Zondervan, 1998); C. L. Blomberg, *The Historical Reliability of the Gospels* (Downers Grove: Intervarsity, 1987); G. R. Habermas, *The Historical Jesus: Ancient Evidence for the Life of Christ* (New York: College Press, 1996).

3.   F. F. Bruce, *The New Testament Documents, Are They Reliable?* (Grand Rapids: Eerdmans Publishing Co., 2003).

4.   Lewis, *Mere Christianity,* 45.

5.   A. Einstein, "Science, Philosophy and Religion: A Symposium" (1941).

6.   J. Polkinghorne, *Belief in God in an Age of Science* (New Haven: Yale University Press, 1998), 18–19.

7.   Copernicus quoted in D. G. Frank, "A Credible Faith," *Perspectives in Science and Christian Faith* 46 (1996): 254–55. Copernican scholar Owen Gingerich has expressed doubts in the authenticity of this quote.

Appendix

1. A more detailed description of the experiences of Susan and her family can be found in M. Waldholz, *Curing Cancer* (New York: Simon & Schuster, 1997), chapters 2–5.

2. T. L. Beauchamp and J. F. Childress, *Principles of Biomedical Ethics,* 4th ed. (New York: Oxford University Press, 1994).

3. D. L. Hamer, *The God Gene* (New York: Doubleday, 2004).

ACKNOWLEDGMENTS

Woodrow Wilson once quipped, "I not only use all the brains that I have, but all that I can borrow." That has certainly been true for me in assembling the ideas and concepts that make up this book. Although I have employed the context of modern studies of the human genome to provide a fresh examination of the potential harmony between the scientific and spiritual worldviews, few if any original theological concepts are portrayed within these pages. I am therefore deeply indebted to a long line of great thinkers from Saint Paul to Saint Augustine to C. S. Lewis, whose ability to discern spiritual truth dwarfs anything that I could imagine producing on my own.

The urge to write this book has been gradually coming into focus for more than two decades, but it took the encouragement

281

of sincere friends to make it a reality. Among the many who have played the role of Barnabas at various times are fellow scientist-believer Dr. Jeffrey Trent; the leaders of the C. S. Lewis Institute Fellows program, the Reverend Tom Tarrants and Dr. Art Lindsley; and my friend and distinguished scholar of C. S. Lewis and Sigmund Freud, Dr. Armand Nicholi. I have also much benefited from the thoughtful writings of other biologist-believers, especially Drs. Darrel Falk, Alister McGrath, and Kenneth Miller.

A particularly important moment in the formulation of the concepts described here was the opportunity to present the Noble Lectures at Harvard in February 2003. On three consecutive evenings at Harvard Memorial Church, I discussed the interface between science and faith, and the turnout of hundreds of Harvard undergraduates each evening convinced me of the hunger that many young people have for discussions of this topic. I particularly thank the Reverend Peter Gomes for making that occasion possible.

Many others have assisted in the birthing process of this book: Judy Hutchinson faithfully transcribed my dictated drafts, Michael Hagelberg nicely rendered the drawings from my sketches, and important critiques of early drafts of chapters were provided by Drs. Frank Albrecht, Ewan Birney, Eric Lander, and Bill Phillips. As my agent, Gail Ross has provided the practical experience that this novice writer sorely needed, and Bruce Nichols has been a perfect editor—providing encouragement for the possibility of this book before I was convinced it could happen, expressing confidence over the rough spots, and setting high standards for clarity and accessibility.

Finally, I thank my family. My daughters Margaret Collins-Hill and Elizabeth Fraker and their husbands consistently offered much-needed encouragement for this project. Still intellectually vibrant in their nineties, my parents Fletcher and Margaret Collins provided critical input to the original plans for this book, though sadly my father did not live to see it come to fruition. I hope he is enjoying reading it from his current address, though I am sure he will identify many unnecessary adverbs that should have been subjected to better editing. Most especially I thank my wife Diane Baker for believing in the importance of this work, and backing that up with countless hours at the computer entering never-ending rounds of edits.

INDEX

ABOUT THE AUTHOR

Francis S. Collins is one of the country's leading geneticists, and longtime leader of the Human Genome Project. Born and raised on a ninety-five-acre farm with no indoor plumbing, Collins grew up an agnostic, then became a committed atheist while getting his Ph.D. in chemistry. It wasn't until he attended medical school and witnessed the true power of religious faith among his patients that his worldview began to change. As a medical geneticist at the University of Michigan, he helped discover the genetic misspellings that cause cystic fibrosis, neurofibromatosis, and Huntington's disease. As head of the highly successful Human Genome Project, he has coordinated the work of thousands of geneticists in six countries. In his spare time, he plays guitar, rides a motorcycle, and writes new lyrics to familiar tunes to entertain his colleagues.

THE
LANGUAGE
OF GOD

A Scientist Presents Evidence for Belief

FRANCIS S. COLLINS

Discussion Group Guide

ABOUT THIS GUIDE

The following author interview and reading group guide are intended to help you find interesting and rewarding approaches to your reading of *The Language of God*. We hope these elements enhance your enjoyment and appreciation of the book.

DISCUSSION GROUP GUIDE FOR

THE LANGUAGE OF GOD

Discussion Questions

1. "So here is the central question of this book: In this modern era of cosmology, evolution, and the human genome, is there still the possibility of a richly satisfying harmony between the scientific and spiritual worldviews?" (pp. 5–6). What view on the integration of science and religion did you have before reading this book? How would you answer Collins's question now?

2. On page 23, Collins sums up the Moral Law, stating that "the concept of right and wrong appears to be universal among all members of the human species (though its application may result in wildly different outcomes)." Do you believe the Moral Law exists?

3. What caused the author to question his atheism? At the

end of the book, he calls on the reader to question his or her current beliefs. Do you think this is a realistic request, or will the average reader wait for a "personal crisis" before embarking on a journey of spiritual discovery (p. 233)?

4. Did the book fairly assess the different religious "options" of atheism, agnosticism, creationism, intelligent design, and theistic evolution, renamed as BioLogos (pp. 159–211)? Did reading these descriptions change your understanding of any of these views? Which option best explains your beliefs?

5. Collins argues that atheism is the least rational of all these choices, since an atheist must claim such extensive knowledge that she or he can conclusively discount the possibility of God. Along those same lines, G. K. Chesterton called atheism "the most daring of all dogmas . . . for it is the assertion of a universal negative." Do you agree? Is it possible to be a rational atheist?

6. Collins states his belief that Young Earth Creationist opinions ultimately harm the religion they represent more than help it: "But it is not science that suffers most here. Young Earth Creationism does even more damage to faith, by demanding that belief in God requires assent to fundamentally flawed claims about the natural world" (p.177). Do you agree?

7. Collins presents data from the study of genomes (pp. 133–141) that argue for a common ancestor of chimps and humans. Do you find the arguments compelling from the anatomy of human chromosome 2, pseudogenes, and ancient repetitive elements? Does common ancestry argue against God?

8. Collins quotes Saint Augustine's warning (in 400 AD) that narrow interpretations of biblical passages with uncertain meaning may place faith at the risk of ridicule if future discoveries conflict with those narrow interpretations (p. 83). In what situations today do you think that warning may have relevance?

9. Discuss the following quote from Galileo: "I do not feel obliged to believe that the same God who has endowed us with sense, reason, and intellect has intended us to forgo their use" (p. 158). What was Galileo trying to say? Do you think this statement is in tune with Collins's views?

10. In the following quote from the book, Collins refers to "why" questions as those for which science is poorly suited to provide answers: "And as seekers, we may well discover from science many interesting answers to the question 'How does life work?' What we cannot discover, through science alone, are the answers to the questions 'Why is there life anyway?' and 'Why am I here?'" (p. 88). Does Collins support this claim elsewhere in the book? Do you agree with him?

11. How does the theme of this book fit together with the opening lines of Psalm 19: "The heavens declare the glory of God; the skies proclaim the work of his hands"?

12. Collins frequently describes the danger of basing religious beliefs on the scientific information that we don't know, referred to as "God of the gaps" (p. 93). "Faith that places God in the gaps of current understanding about the natural

world may be headed for crisis if advances in science subsequently fill those gaps" (p. 93). However, he also says that the answers he searches for are those that science alone cannot discover (p. 88). Does Collins's personal search fall within his description of looking for God of the gaps? Why or why not? See pages 193 and 204 for more references to God-of-the-gaps thinking.

13. Do you foresee a time when all religious leaders will accept Darwinism, just as we eventually came to accept that the earth revolves around the sun? Is the battle between science and religion destined to continue over each new scientific discovery that is made?

Enhance Your Book Club or Discussion Group

1. In November 2006, *Time* magazine hosted a debate between Francis Collins and Richard Dawkins, author of *The God Delusion,* for a cover story (see http://www. time.com/time/magazine/article/0,9171,1555132,00.html). Bring it to your group, and discuss the strengths and weaknesses of the arguments.

2. Reread Genesis 1 and 2, or read these passages for the first time, or bring them to your meeting to discuss with your group. Do you see the two slightly different creation stories? How do you interpret these verses now that you've read *The Language of God*?

3. C. S. Lewis is quoted frequently by Collins as the philosopher who helped him discover God, and Collins repeatedly

quotes Lewis's work when important religious questions arise. Take a trip straight to the source and read one of the Lewis books that Collins quotes. Choose from among *The Problem of Pain, Surprised by Joy, Mere Christianity,* and *Miracles.* For more information on C. S. Lewis, visit www.cslewis.org.

4. As the head of the Human Genome Project, Collins has attracted a great amount of attention in the press and on the web. Search the Internet for the information that interests you most about Collins, and print a copy to bring in for discussion with the group. Good places to start your search include www.genome.gov, www.cnn.com, www.salon.com, and www.nytimes.com. On the website for the PBS series *Religion and Ethics,* you can watch an online video of Collins at www.pbs.org/wnet/religionandethics/week947/profile.html.

5. Take your online research of Collins a step further to discover sites dedicated to contemplating the coexistence of science and religion. Head to the website of the C. S. Lewis Foundation at www.cslewis.org or take a look at the companion site of the four-hour PBS special *The Question of God:* www.pbs.org/wgbh/questionofgod. Visit www.god andscience.org, www.hawking.org.uk, www.ucmp.berkeley.edu/history/thuxley.html, and www.aboutdarwin.com for even more information.

6. Try finding websites that explain more about the scientific and medical topics mentioned in the book. Search for de-

tails on the Big Bang, cystic fibrosis, personalized medicine, and the human genome. Sites such as www.umich.edu/~gs265/bigbang.htm, www.cff.org, www.personalized medicinecoalition.org/, and www.genome.gov are good places to start.